PHILIP'S

ESSENTIAL GUIDE TO
SPACE

PHILIP'S

ESSENTIAL GUIDE TO
SPACE

THE DEFINITIVE GUIDE TO EXPLORING AND UNDERSTANDING OUR SOLAR SYSTEM AND THE UNIVERSE BEYOND

Paul Sutherland

For Terry, who offered much encouragement when I began this project but sadly did not live to see it completed. Thanks for all the banter, Old-timer!

Paul Sutherland was five years old when the Soviet satellite Sputnik went into orbit. He recalls the excitement of Apollo and staying up all night to follow the first Moon landing. He spent much of his career working as a subeditor on a number of Fleet Street titles, before going freelance to focus on communicating space science in plain English. Paul has since contributed regularly to magazines including *Astronomy Now*, *Sky at Night* and *Focus*, and space website Sen.com, and has written for newspapers from *The Times* to *The Sun*, which dubbed him 'The Sun Spaceman'. He has actively supported the Society for Popular Astronomy, including editing its publications. And out in the depths of space lurks an asteroid officially named for him, and carrying his Twitter handle, Suthers.

Designer Chris Bell, cbdesign

www.philipsastronomy.com
www.philips-maps.co.uk

First published in Great Britain in 2016 by Philip's,
a division of Octopus Publishing Group Limited
(www.octopusbooks.co.uk)
Carmelite House, 50 Victoria Embankment,
London EC4Y 0DZ
An Hachette UK Company (www.hachette.co.uk)

ISBN 978–1–84907–419–3

A CIP catalogue record for this book is available from the British Library.

Printed in China

CONTENTS

CONTENTS

VOYAGES OF DISCOVERY

Throughout history, humankind has been confined to a tiny speck in a vast Universe. But in the last century we have broken free from this cosmic island to send unmanned vessels to other worlds that circle our local star, the Sun. Robotic emissaries have since travelled as far as the edge of the Solar System. Hundreds of astronauts have now orbited the Earth, and a tiny handful have even walked on our closest neighbour, the Moon.

Space touches all our lives. Those left on Earth benefit hugely from exploration of the final frontier. Miniaturization needed to send experiments into orbit and beyond has boosted technology to give us smartphones we can slip in our pockets, slim laptops and other electronic devices. Global positioning satellites provide the satnav that sends drivers to the correct destination or helps a busy traveller find a nearby restaurant. Satellites support worldwide communications, such as the internet, news reporting or live viewing of major TV sporting events. Others forecast the weather, detect forest fires, aid farming and support disaster responses.

Experiments aboard the International Space Station are giving hospitals on Earth advanced medical techniques, including scanning for breast cancer, treatment of the bone disease osteoporosis, and portable devices to detect tuberculosis and other conditions in remote parts of the world.

But practical benefits aside, a driving force has been humanity's timeless urge to explore. Early people wondered what lay over the next hill, or beyond the horizon. Humankind expanded far beyond its early settlements in north-east Africa and around the world. Mountain ranges, deserts and rivers proved but minor obstacles, and even that vast and moody barrier, the sea, could not hold explorers back.

Today's manned missions are like the great voyages of exploration made across the world's oceans in the 17th and 18th centuries, when courageous crews set sail in the face of unknown challenges, dangerous currents and the threat of violent storms. Thousands perished in the drive to advance human knowledge about the world. Newly discovered lands brought great rewards, such as resources to plunder. However, the prime objective was surely to satisfy an innate curiosity about what lies out there, beyond familiar surroundings.

The cold, dark vacuum of space represents a new 'ocean' that presents fresh risks and perils. Ships' masts have been replaced by antennas and solar panels, sails by new forms of propulsion, and wooden cabins by pressure-sealed capsules. But the urge to discover what lies out there remains the same as ever, and voyagers to the new horizons will show similar bravery and daring to their maritime predecessors.

This book offers a summary of humanity's progress so far in reaching and exploring space, what we have learned about our neighbouring planets and more remote objects in the Universe, plus a look to future missions to the unknown. I hope it helps you to discover some of the excitement and wonder of space!

Paul Sutherland

◀ A colourful cloud of gas and dust in the constellation of Camelopardalis is revealed by NASA's thermal-imaging satellite WISE to be the incubator for a family of infant stars.

INTRODUCING THE UNIVERSE

An immense city of stars shines as a whirlpool of light in the heavens. It is nearly a century since astronomers recognized that some nebulous glows were really collections of billions of suns called galaxies, like islands in a vast cosmos. The Hubble Space Telescope has shown many in great detail, and countless others near the beginning of time, helping us learn more about the Universe and how it began.

OUR PLACE IN SPACE

We live in a tiny part of a vast Universe. Our Sun is one star among a couple of hundred billion stars in the Milky Way Galaxy. And that galaxy is just one among hundreds of billions of galaxies in the Universe. Everything is on an unimaginable scale. But telescopes on Earth and satellites in space are helping us understand more about what is in it and how it came to exist.

HOW THE UNIVERSE BEGAN

Astronomers' evidence points to everything, including space and time, simply bursting into existence 13.8 billion years ago. This awesome event, which we call the Big Bang, saw all space, matter and energy appear as an incredibly hot, dense point, smaller than an atom. It was a microscopic fireball that grew a trillion times bigger in just a trillionth of a second! Within one millionth of a second, the Universe had expanded to a diameter of around 1 billion km, and the temperature dropped from 10 billion trillion trillion degrees Celsius to 10 trillion degrees – still hot but considerably cooler. From this energetic start emerged a soup of particles and anti-particles which reacted to produce the first protons, neutrons and other basic constituents of matter.

After 100 seconds, with the Universe now many hundreds of light years across, the nuclei had formed of the lightest elements' atoms, mainly helium plus hydrogen.

There would have been nothing to see if you had been there, because the Universe was then a completely opaque fog. Photons that make up light were being scattered about but unable to shine. Then about 300,000 years after the Big Bang, the cosmic fog began to lift. Those atomic nuclei were able to capture electrons to make complete atoms. Clouds of hydrogen and helium gas formed. Photons were set free from the rest of matter and the Universe became transparent.

Hundreds of million years more had to pass – a period that astronomers term the Dark Ages – before stars and galaxies formed and began to shine, producing the Universe's heavier elements as they did so.

▼ Artificial colours indicate temperature fluctuations in this whole-sky map of the early Universe built up by the WMAP probe over seven years.

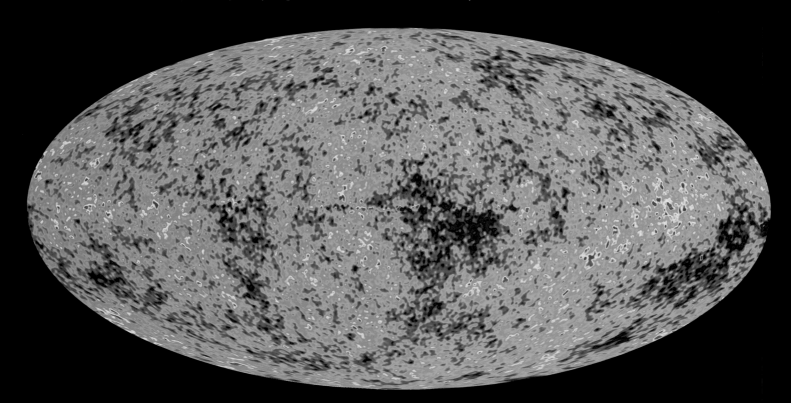

▶ A 3D representation of the timeline from the Big Bang to the present, indicating how the Universe is now expanding more rapidly.

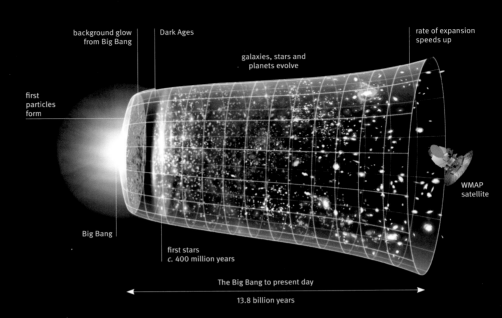

background glow from Big Bang

Dark Ages

galaxies, stars and planets evolve

rate of expansion speeds up

first particles form

Big Bang

first stars c. 400 million years

WMAP satellite

The Big Bang to present day

13.8 billion years

At the start of the Space Age, there was still support for a rival idea, which was that the Universe had always existed and was in a 'steady state'. Indeed, the term 'Big Bang' was thought up by a leading UK astronomer, Fred Hoyle, as one of derision. Astronomical observations, however, increasingly showed that the event really did happen. In the 1920s, fuzzy patches in the heavens had been identified as being other galaxies outside our own Milky Way. Surveys of them by Edwin Hubble with a 100-inch telescope on Mount Wilson in California showed they were moving away from us, and the farther away they were, the faster they were receding. The Universe was expanding.

Predictions suggested that a glow from the Universe's fiery birth would still be detectable as background radiation. It was heard for the first time in 1964 by radio astronomers Robert Wilson and Arno Penzias using an antenna at Holmdel, New Jersey. Satellites helped confirm the glow's existence. NASA's Cosmic Background Explorer satellite (COBE) was launched in 1989 to measure the diffuse infrared and microwave radiation from the early Universe, and found that observations closely matched predictions of the Big Bang theory.

In 2001, NASA's Wilkinson Microwave Anisotropy Probe (WMAP) was launched to a point 1.5 million km away in space to probe conditions after the Big Bang. It began to map the Cosmic Microwave Background in greater detail, and was joined in the task by Europe's Planck Space Telescope, named after leading German theoretical physicist Max Planck, which launched in 2009.

Astronomers who examine the nature and origins of the Universe are known as cosmologists. Some have suggested that our Universe might be one of many – a concept termed the Multiverse. It has been suggested that 'bruises' seen in the Cosmic Microwave Background could be due to collisions with other universes, though the idea is highly speculative and controversial.

MYSTERIOUS DARK FORCES

There is still much that is not understood about the Universe. One problem is that it seems we can only see a small fraction of the material in space and that up to ten times as much is an invisible substance called dark matter. Though it has never been directly detected, its presence is inferred by its effect on stars and galaxies.

Another major mystery is a powerful force known as dark energy. The force is believed to account for 75% of the energy-mass of the Universe and to explain why its expansion is speeding up and not slowing down.

HOW IT WILL END

No one can say for sure how the Universe will end. One suggestion has been that it will eventually fall back in on itself in a Big Crunch. But the fact that the rate of its expansion is accelerating suggests everything will eventually be so cool and dispersed that it becomes a dark and empty void.

◀ Thousands of galaxies are revealed in a tiny patch of sky by the Hubble Space Telescope. Some are seen as they were when the Universe was just 800 million years old.

THE MILKY WAY GALAXY – A CELESTIAL WHIRLPOOL

Of the countless billions of galaxies in the Universe, one is very special for us because it is home. We call it the Milky Way Galaxy as, seen from within, it appears like a bright milky band across the night sky. Our galaxy is of a typical spiral type – a flat rotating disk, 100,000 light years wide, with a central bulge. Our star is located a little over halfway out from the centre in one of its spiral arms, where the galaxy is only 2,000 light years thick.

Strangely we know less about the shape of our home galaxy than those at remote distances. That is because we are trying to study it from within, so it is rather a case of not seeing the wood for the trees. However, the Sun is thought to lie 27,000 light years from the galaxy's centre in one of its four main spiral arms. These arms, made up of individual streams of stars, are formed by density waves where regions of star formation become bunched together, rather like a cosmic traffic jam.

The Sun is orbiting within the Milky Way and takes 220 million years to make one trip around the centre, travelling at 828,000 km/h. It has completed 20 orbits since it was born.

It is believed that our galaxy is of a type called a barred spiral, having a bar formation at the middle, but no one is absolutely sure. One problem is that vast clouds of dust block the centre of the galaxy from view, so only telescopes observing infrared light (heat) or radio telescopes are able to penetrate this veil. At the heart of the galaxy there lies a supermassive black hole, as is common with galaxies like our own, with a mass a billion times that of the Sun. It steadily grows in size as it draws in stars and dust that approach too closely.

▼ How our Milky Way spiral galaxy might appear if we could view it from outside and above, revealing the central bar.

Occasionally a very dense, high-mass star in the Milky Way will explode and destroy itself in a supernova when it comes to the end of its much shorter life. The energy produced can be as much as all the other stars in the galaxy. Remnants of supernovae that exploded centuries ago can be seen in the night sky. The last to be seen explode in our neighbourhood was in 1987 in a small companion to the Milky Way, the Large Magellanic Cloud. It was a gift to astronomers as it allowed them to study the process in detail for the first time with space telescopes.

Supernova explosions provide the material for new star systems to form. Ancient supernovae also produced the rare and precious metals such as gold and silver that we find on Earth today.

THE COSMIC ZOO

A nebula is a cloud of gas and dust, often illuminated by starlight. Some, such as the Great Nebula in Orion, are stellar nurseries where new star systems are being born.

The Milky Way Galaxy is surrounded by a halo of densely packed balls of many thousands or millions of stars, called globular clusters, rather like bees buzzing around a hive. These contain some of the oldest stars in the Universe. They have survived a long time, maybe 12 billion years, because they are low in mass.

An open cluster is a collection of stars that formed together. A famous example is the Pleiades, or Seven Sisters, in the constellation Taurus.

All the stars we see with the naked eye in our night sky are in our own Milky Way Galaxy. Patterns of stars gave ancients the constellations, but stars in constellations are not physically linked. They simply lie in the same general direction and can be at hugely varying distances.

▲ This spiral galaxy, labelled NGC 3949 and found in the constellation of Ursa Major, is thought to resemble our own Milky Way star system.

CLUSTERS OF GALAXIES

The expansion of the Universe tells us that objects within it are all moving away from its centre and each other. But on a local scale, gravity causes galaxies and other bodies to interact with each other. Many galaxies exist in clusters or even superclusters. Our own Milky Way and its bigger neighbour M31, the Andromeda Galaxy, 2.5 million light years away, are actually closing together and will collide in 4 billion years' time. They are part of what is called the Local Group, made up of at least 46 galaxies, including small, less dense stragglers known as dwarf galaxies.

▼ An impression of how our Milky Way will be distorted by its collision with neighbouring galaxy M31 in 4 billion years' time.

OUR HOME – THE SOLAR SYSTEM

Our immediate home in the Universe is, of course, the Earth – one of eight planets that orbit the Sun. Another outer world, Pluto, was once also labelled a planet, but was relegated to a lower league in 2006 by astronomical authorities, becoming a dwarf planet instead. Together the Sun, planets and other bodies in its orbit, such as asteroids, are known as the Solar System.

In the last 20 years, astronomers have discovered that other stars have planetary systems too, with thousands of so-called exoplanets being discovered around them by advanced telescopes both on Earth and in space.

A STAR IS BORN

We now know that the Sun and its family of worlds all formed around 4.6 billion years ago in a vast cloud of gas and dust within the Milky Way Galaxy. As the cloud condensed, its material fell together to form many stars, including the Sun at the centre of a wide, spinning dusty disk. The fragments of dust themselves came together to form disks of material called proplyds. These planet-building ingredients then combined over many millions of years to produce objects called planetesimals, which then merged to make protoplanets. As this violent early phase continued, the protoplanets ran into each other, like a cosmic version of interplanetary snooker, until we were left with the planets that we know today. Any remains of the original dusty disk were swept away by the force of the solar wind, a breeze flowing from the Sun.

▼ A highly stylized representation of the Sun and the eight planets now recognized within our own Solar System. The diagram is not to scale.

MEET THE FAMILY

The inner planets of the Solar System – Mercury, Venus, Earth and Mars – are all rocky worlds. They are followed by a belt of countless rocky fragments called the asteroids, believed to be fragments left over from the start of the Solar System that were unable to combine to form a planet, due to mighty Jupiter's disruptive influence.

Beyond the asteroids come the giant planets Jupiter, Saturn, Uranus and Neptune. These are quite different from the inner worlds, being largely composed of gas or ice.

The reason for these different characteristics is the intense heat in the early inner Solar System, which was approaching 2,000°C, compared to a cool −200°C in the outer reaches. Only elements with high melting points survived in the inner zone to form the rocky bodies, and the lighter ones were vaporized. Farther out, beyond a point known as the snow line, or frost line, substances such as water and methane remained intact and collected around small cores of rock and ice to form the giant planets. They also gathered large quantities of hydrogen and helium in the process.

Clearly there is water on the Earth today. More than 70% of the planet is covered with it. Planetary scientists believe that this water was either delivered later by colliding comets, or was trapped within the rocks that formed Earth. No one is yet sure what the major provider was.

Ex-planet Pluto marks the start of an outer zone of numerous smaller icy bodies known as the Kuiper Belt. And more than a thousand times farther out, and perhaps stretching halfway to the nearest star, is believed to exist a sphere of icy fragments called the Oort Cloud. This vast reservoir is the source of many comets that got deflected and headed into the inner Solar System.

▲ A cluster of young stars shining in a celestial nursery in the constellation of Carina, imaged by the Hubble Space Telescope. The Sun formed in a similar cloud of gas and dust.

NEW PLANET NINE?

Early in 2016, astronomers studying the orbits of objects in the Kuiper Belt claimed to have found compelling evidence for a ninth planet in the Solar System, similar in size to Neptune. The object has yet to be identified, but the team say it is controlling the behaviour of a number of small icy bodies. They suggest it follows a highly elliptical orbit at an average distance of 90 billion km, and taking between 10,000 and 20,000 years to make one circuit of the Sun.

▲ If it exists, Planet Nine could resemble this artist's impression of a world on the edge of the Solar System.

LEAVING THE EARTH

NASA's Space Shuttle *Discovery* blasts off from Florida and into orbit around the Earth. Rocket launches to space stations, the Moon and planets have become almost routine, but we owe such frequent flights to the philosophers, scientists and engineers who first dared to imagine how humankind might one day break the bonds that tied us to the Earth and explore a whole new frontier. And then made it happen.

PIONEERS IN ROCKETRY

The earliest rockets were probably created by the Chinese around 2,000 years ago. They made fireworks, using gunpowder as an explosive, and some must have been sent flying by the force of the blast. Soon they realized that they could deliberately fill tubes with gunpowder and the thrust generated would propel the tubes some distance. Typically for humankind, the discovery was used to make weapons, and the rockets became an advanced form of arrows in wars between different tribes. Thousands of years later, the same basic principle is used today to give us the rockets that light up the sky in colourful celebratory firework displays.

News of this advanced form of weaponry crossed the world and similar rockets were used by armies in Europe between the 13th and 15th centuries. In these early times, some philosophical types imagined leaving the Earth's surface and journeying into space. Usually these ideas were purely fanciful, involving being launched by powerful springs, or in chariots with feathered wings. But legend has it that in the 16th century a Chinese mandarin named Wan Hu built a 'spacecraft' by attaching 47 rockets to a chair fitted with kites. A team of assistants lit the fuses before running for cover – after the explosion that followed, nothing was seen of Wan Hu or his flying chair ever again.

This story may well be apocryphal, but in the 17th century the French dramatist Cyrano de Bergerac imagined advanced forms of propulsion, including the rocket. And two centuries later, in 1865, his fellow countryman Jules Verne wrote the novel *From The Earth To The Moon* and its sequel *Around The Moon*, one of the earliest serious attempts at science fiction. Though it included such fantastic ideas as being fired into space by a cannon, Verne's crew of three launched from Florida, reached

▶ Robert H. Goddard, who led rocket development in the United States, is pictured working on one of his early experiments.

the Moon, then returned to splash down in the Pacific Ocean. Around 100 years later, the Apollo astronauts travelled a similar route for real!

FATHERS OF SPACEFLIGHT

While the novelists had the imagination to dream, other minds came to examine the reality of the science behind space exploration. A key figure in developing ideas behind rocket propulsion, in 1898, was Russian schoolteacher Konstantin Tsiolkovsky (1857–1935). He produced a report that suggested liquid fuel would give rockets their greatest power, and that a rocket's speed and range were only limited by the force of the thrust from the escaping gases. Tsiolkovsky calculated the force that would be needed to leave Earth's grip, now known as 'escape velocity', and the speed necessary to keep a spacecraft in orbit.

Tsiolkovsky is considered the father of modern spaceflight, though his work was purely theoretical and he did not build any rockets. But working independently on the other side of the world was an American scientist, Robert H. Goddard (1882–1945), who carried out his own experiments, including building rockets, first powered by solid fuels and later liquid propellants. His first successful flight came with a liquid-fuelled rocket in March 1926. It climbed only 12.5 m in a flight lasting two and a half seconds, but was a landmark in rocketry. Later rockets flew higher and further, and Goddard used parachutes to bring them back to Earth.

▶▲ Author Jules Verne (right) alongside an illustration from his classic novel *From The Earth to the Moon* (above).

Goddard's genius included designing gyroscopes to keep his rockets stable, and putting vanes in their jet stream to steer them. Today, a major NASA centre in the US state of Maryland is named after him.

A third leading figure in early rocketry was Hermann Oberth (1894–1989), who was born in what is now Romania. He began building rockets as a teenager, and later his work in Germany directly inspired the development of rockets as Nazi weapons at the tail end of World War II. Built with the aid of slave labour from concentration camps, more than 3,000 V-2 missiles were fired into England and Belgium, killing many thousands of civilians.

At the end of the war, leading rocket engineers, plus their technology, were acquired by both America and the Soviet Union, so spurring the development of missiles and the space programmes for both sides.

◀ Konstantin Tsiolkovsky, surrounded by early rocket designs.

HOW ROCKETS WORK

The basic principle behind how a rocket works can be demonstrated simply with a party balloon. Blow it up, then let it go and it will fly off in the opposite direction as a blast of air escapes from the neck. It may not be a very controlled flight, but the balloon is illustrating a physical rule that was first stated as early as 1686 by the great English mathematician Sir Isaac Newton. In the third of three laws identified by Newton, he said: 'For every action, there is an equal and opposite reaction.' This explains how the production of thrust by a rocket engine will propel it in the opposite direction, even in the vacuum of space.

In a rocket, the force is produced by the combustion of fuel in its engine, turning it into hot gas. As this gas, smoke and flames are expelled from the nozzle end towards the ground, the vehicle is propelled in the opposite direction, lifting it skywards. To reach space, a rocket has to build up enough speed to counter the Earth's gravitational pull. Continual controlled burning of the fuel allows the rocket to accelerate. Spacecraft have to reach a speed of 28,400 km/h to go into a simple orbit, but more than 40,000 km/h to escape the Earth's pull altogether.

Rockets achieve the continual acceleration needed to reach Earth orbit or beyond by combining more than one stage, each containing its own reservoir of fuel. Commonly, these are stacked one on top of another, known as serial staging. When the lowest stage's fuel has been exhausted, it is jettisoned and the next stage above takes over. The Saturn V rockets that launched Apollo to the Moon used three stages, as does Russia's veteran Soyuz fleet which is still in use today.

Other rocket designs use a number of rockets strapped together, known as parallel staging. NASA's Space Shuttles were launched using this technique, having a pair of rocket boosters alongside a main giant fuel tank. Europe's main workhorse, the Ariane 5, has a similar configuration.

On top of the rocket is placed the payload being carried into space, for example one or more satellites, or a planetary probe. This special cargo is protected by an aerodynamic fairing, or nose cone, which is jettisoned once above the atmosphere.

Fuels that propel rockets come in two main types – liquid and solid. Either may be used, and some use a combination of both, such as the Space Shuttle and the Ariane 5 again, designed to carry liquid fuel in their main tanks, but launched with pairs of solid rocket boosters attached. Liquid fuels are typically a mix of hydrogen

▶ A Soyuz TMA 10 lifts off from Baikonur, Kazakhstan, in 2007. This veteran of the Russian space fleet is an example of a three-stage rocket.

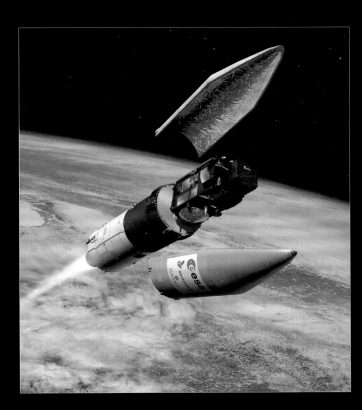

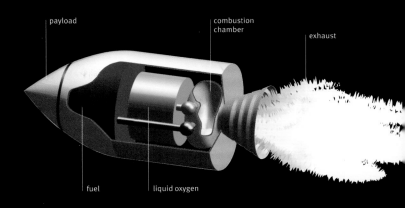

A simple diagram demonstrates the basic principle of a rocket. Fuel, often hydrogen, is mixed with an oxidant, such as liquid oxygen, and burned, producing thrust from the exhaust.

◄ This illustration shows how the fairings protecting a Sentinel-3 Earth-monitoring satellite are discarded by its Russian Rockot launch vehicle as it heads into orbit.

and oxygen, and solid fuel might combine aluminium powder with a supply of oxygen to ignite it.

COMING HOME

The early rocket stages to fall away during a launch commonly drop into the sea and are designed to be used once only, though the Space Shuttle's smaller boosters were recycled, and commercial launch companies are trying to perfect a system where they can return to land and be used again. Upper stages usually go into orbit to add to the many man-made items circling our planet. Sometimes unwanted rocket parts are steered back into the atmosphere to burn up in the heat of re-entry, or will fall out of orbit the same way due to gradual air resistance. Crewed capsules that need to protect their occupants are fitted with insulating heat shields and brought back through the atmosphere in a controlled manner to deliver the astronauts safely back to Earth. Space capsules complete their descent by parachute, either into the sea for American missions or a sparsely inhabited region of land for Russian and Chinese vehicles. Uniquely so far, the Space Shuttle landed on a runway for reuse, but the method is planned for some future commercial spacecraft.

▼ The Space Shuttle *Atlantis* blasts off from Florida, on a rocket stack that combines liquid- and solid-fuelled boosters.

PROPULSION AND POWER

Tanks of rocket fuel have been used to get spacecraft into orbit, but other forms of propulsion are being developed to transport them across the vacuum of space.

Solar sail: This involves unfurling a large, thin surface such as aluminium-coated plastic to catch sunlight. The pressure of this solar radiation produces a gradual force, like a light breeze on a yacht's sails, but which can accelerate a spacecraft to high speeds over time. Both Japan and NASA have tested the technology and NASA is planning to use it to develop a CubeSat (see page 27) that can encounter close-passing asteroids. Some scientists believe the technology could be adapted to allow powerful laser beams on Earth to propel the sails even as far as nearby stars.

Electric: Various types of electric engine take a chemical propellant, then use electrical power to accelerate it, for example by heating it. This has long been a popular idea in science fiction, including powering the *Discovery One* spaceship in Arthur C. Clarke's *2001: A Space Odyssey* – the technique has been used in real life too. Ion engines are a variation, which charge a gas such as xenon in a magnetic chamber to produce thrust from a flow of positively charged atomic particles. NASA's Dawn mission to asteroids Vesta and Ceres was powered by ion thrusters.

Gravity assist: Whatever onboard fuel is used, a number of missions to deep space have used the pull of the planets to help propel them. Instead of flying directly to a target, a probe will be slung around the Solar System to gain or lose momentum.

▼ An artist's view of NASA's Juno probe getting a boost from Earth's gravity as it flies past on its way to giant planet Jupiter.

▲ A concept image for a solar sail which would use light to accelerate a spacecraft across space. Laser-driven sails might allow small robotic craft to reach nearby stars.

POWERING SPACECRAFT

Solar arrays: Early satellites were powered by simple batteries, but these soon ran out, causing the spacecraft to stop functioning. Today, most spacecraft have solar arrays as their main power supply, to collect sunlight and turn it into electrical current. The first use of this technology was in 1958, in order to power NASA's Vanguard satellite. Such power is stored in rechargeable batteries for when the satellites lose sunlight, for example on the night side of Earth. Solar power is especially useful for spacecraft in the inner Solar System because farther out the supply of sunlight is considerably reduced. Missions such as the European Space Agency's Rosetta comet explorer carried huge solar arrays to compensate, resembling widely outstretched wings.

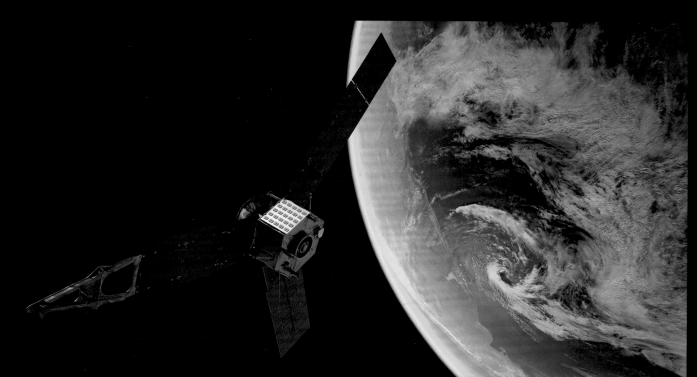

▲ NASA's New Horizons probe, *en route* to Pluto, passes Jupiter in 2007 to get a speed boost from the giant planet's gravity in this artist's impression.

Nuclear generators: Missions to the outer Solar System, where sunlight is not sufficient, have carried generators powered by radioactive material such as plutonium. This has been controversial, but the fuel's containers are designed to prevent any contamination even in the event of a launch failure. The power systems, termed radioisotope thermoelectric generators, or RTGs, have allowed the Pioneer and Voyager probes to tour the most remote worlds in our Solar System, and Galileo and Cassini to spend years orbiting Jupiter and Saturn. They were also used on the Apollo Moon missions, and to power NASA's Viking Mars landers as well as the Curiosity rover that is still exploring the Red Planet.

STAYING UPRIGHT IN SPACE

Spacecraft have to avoid tumbling in space and a variety of methods control their orientation. A simple way is to spin, and this has been used to stabilize many satellites, just as the spinning wheels of a bicycle give it stability. However, it does not suit spacecraft that are intended to be held in a fixed arrangement. For these, techniques are used to control their movement along three axes. One way is by using thrusters, but these mini-rockets depend on a supply of fuel. An alternative is to use

individual heavy wheels that spin within the spacecraft, known as reaction wheels, to give stability. NASA's Kepler Space Telescope used four reaction wheels in order to keep pointing at one tiny patch of sky and discover thousands of planets around other stars. When two wheels failed, the mission had to be radically reimagined. Some satellites, including the Hubble Space Telescope, use electromagnets to align themselves with the aid of the Earth's magnetic field. Spacecraft voyaging across the Solar System use sensors to track the Sun and bright stars to help keep them oriented correctly.

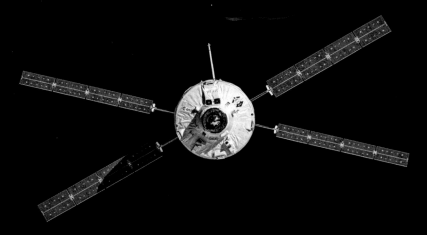

▲ The European ATV cargo ship *Albert Einstein* shows its four large solar panel arrays as it approaches to dock with the International Space Station.

SATELLITES AND SPACECRAFT

There is no sudden boundary between the Earth's atmosphere and outer space. However, the air becomes rapidly more rarefied as one travels further above the surface, and an altitude of 100 km has been declared the point where space begins. It is known as the Kármán line after the engineer and physicist Theodore von Kármán (1881–1963), who calculated that no aircraft could fly at this height or beyond.

CIRCLING THE GLOBE

There are currently around 2,300 man-made spacecraft orbiting the Earth – and that does not include the myriad of spent rocket stages and chunks of debris that are also up there. These spacecraft may be powered by an onboard battery or solar panels, which convert sunlight into electricity. Satellites that are still functioning today focus on the Earth as much as space, using cameras and other sensors, and include the following types.

Military: Sometimes labelled spy satellites, these are launched by major world powers to keep an eye on other countries' affairs, provide intelligence, and support military services on the ground.

Communications: An important part of the infrastructure that supports modern living, these route phone calls, beam TV signals, support the internet, and allow virtually instant contact between remote regions of the world by relaying signals received from ground stations.

▶ The first image of the Earth taken by a second-generation Meteosat satellite in colour in January 2006.

▶ An illustration showing a European Meteosat weather satellite separating from the upper stage of its Ariane 5 launcher to enter orbit around the Earth.

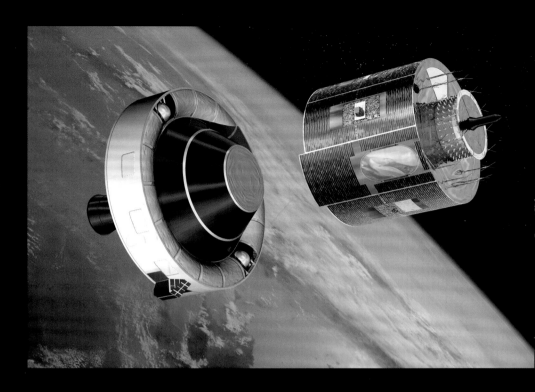

Surveys: These are a vital tool today to help humans track weather systems including typhoons, monitor air quality and study climate change from its effects on ice sheets, the oceans and atmosphere, observe land use to help crops to grow, and in disaster relief to assist in the recovery from catastrophic incidents.

Global positioning: A network of satellites around the globe enables the precise determination of locations required for the satellite navigation (satnav) we use in our cars or to pinpoint the positions of our mobile phones or cellphones.

Scientific: Some satellites are forms of telescope put above the disturbing effects of the atmosphere to observe the Universe in many different wavelengths of light. Others perform studies of Earth, such as the variations in gravitational pull across the planet.

Crewed: Some spacecraft that spend time temporarily in orbit are there, of course, to send astronauts on missions, including especially to the International Space Station (ISS). Similar vehicles fly unmanned to deliver supplies.

Space stations: Habitable environments placed in orbit to support continued occupation by astronauts. The best known is the International Space Station, which has been permanently crewed since November 2000.

CubeSats: Miniature satellites that can be built at low cost by companies or universities, often using off-the-shelf parts. Several may be launched together as secondary payloads to accompany a main spacecraft.

TYPES OF ORBIT

Though we think of satellites circling the Earth, their orbits are generally eccentric, or elongated. The closer to the planet a spacecraft orbits, the faster it has to travel to avoid being pulled to the surface by the force of gravity.

Many spacecraft fly in what is called a **low-Earth orbit** (LEO) at a height of between 160 km and 2,000 km. The International Space Station orbits at an altitude of around 400 km at a speed of 28,000 km/h. The ISS and many other satellites fly in an orbit inclined to the equator so that, over

time, they fly over much of the planet's surface. But some satellites fly in **polar orbits**, literally passing over or close to the north and south poles. The Earth rotates beneath them as they travel, allowing them to survey the whole planet over time.

Communications satellites usually have to fly in what is called a **geostationary orbit** that allows them, as the name suggests, to remain over a fixed point on Earth. This enables satellite TV dishes, for example, to remain fixed yet keep pointing at them. This orbit can only be achieved above the equator and at an extremely high altitude of 35,786 km, with the satellite orbiting at a speed of more than 11,000 km/h. Other types of orbit that are tied to the Earth's own rotation period are called **geosynchronous orbits**. A **transfer orbit** is the trajectory a spacecraft takes to move from one orbit to another.

▲ A number of CubeSats are released into space to carry out low-cost experiments in orbit in this artist's impression.

BEYOND EARTH'S ORBIT

Many spacecraft have been sent beyond the orbit of Earth, including to the Moon and planets. They include the pioneering Apollo missions that landed the first humans on the Moon, but otherwise have been entirely robotic so far. There are a variety of types.

Flyby: Several early space missions launched by the competing space nations of the time, the Americans and the Soviets, were aimed to fly past their intended target, whether the Moon or a planet, gathering information in the process. A flyby by Luna 3 in 1959 gave us our first view of the far side of the Moon, and in 1964 NASA's Mariner 4 sent home the first close-up pictures of Mars. NASA's Pioneer 10 and 11 and Voyager 1 and 2 spacecraft flew past a number of outer planets. Some planetary flybys are made by spacecraft today to gain speed and send them on to a different ultimate destination. In July 2015, NASA's New Horizons probe made a high-speed flyby of Pluto that gave us our first detailed views of this remote world and its main moon Charon.

▼ This artist's impression shows NASA's Cassini probe in orbit around ringed planet Saturn, as it studies the giant world and its many moons.

▲ The European Space Agency's Rosetta space probe is shown making a close pass of Mars in February 2007 to get a speed boost on its way to visit Comet 67P/Churyumov–Gerasimenko in this illustration.

Orbiters: As the name suggests, these spacecraft are sent to circle other worlds, usually in order to survey them over a prolonged period of time. Command modules on the Apollo missions orbited the Moon to support the manned landings before bringing the crews home again. Robotic spacecraft are now regularly sent to Mars to study it from orbit, and others have spent years gathering data about

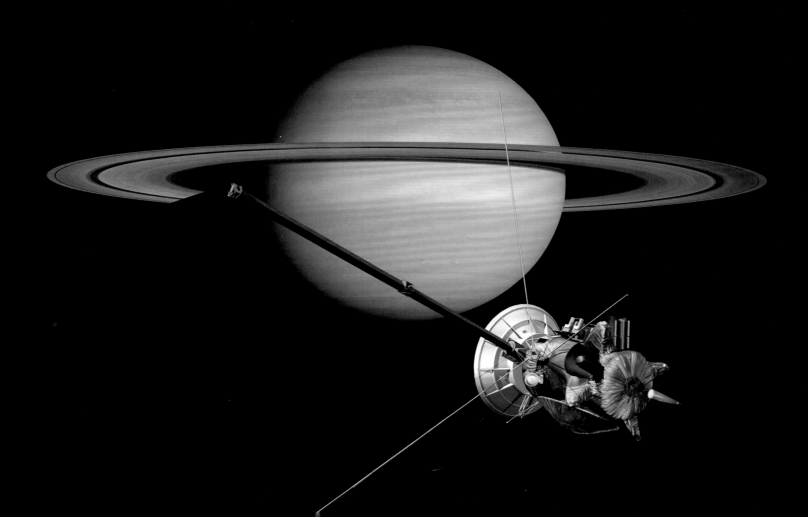

▶ A beautiful view of the Earth as seen by the Apollo 17 crew on 7 December 1972 as they headed towards the Moon to land for the last time on the lunar surface. The coastline of Africa is clearly visible.

the Moon, Mercury, Venus, Jupiter and Saturn. Some spacecraft are sent into orbits around the Sun to monitor its activity.

Landers: Successfully placing a probe on the surface of the Moon or another planet can clearly tell us much about that world. The first probes to 'land' on the Moon, Mars and Venus actually collided in high-speed impacts. Controlled soft landings have followed on the Moon, Venus, Mars, Saturn's largest moon Titan, asteroids, and even a comet called 67P/Churyumov–Gerasimenko. The Apollo lunar modules and some Soviet robotic landers were able to launch again to return NASA astronauts and samples of moonrock to Earth.

LAGRANGIAN POINTS

There are five distant locations in space where a spacecraft can sit in a stable and harmonious relationship with the Earth and the Sun. They are termed the Lagrangian or L points, after 18th-century Italian mathematician Joseph-Louis Lagrange. The L_1 point allows a satellite to sit permanently between the Earth and Sun, an ideal location for the SOHO spacecraft to monitor the Sun and any 'space weather' heading our way. The L_2 point keeps a spacecraft on the far side of the Earth from the Sun, and has been picked for a number of orbiting observatories, including the upcoming James Webb Space Telescope. Both the L_1 and L_2 points lie about 1.5 million km from the Earth. L_3 is the point in the Earth's orbit on the far side of the Sun, and L_4 and L_5 lie 60 degrees ahead and behind the Earth in its orbital path.

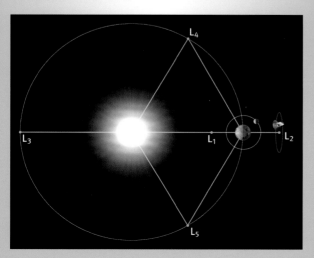

▲ NASA's latest Mars rover, Curiosity, created this 'selfie' by taking many images with a camera on the end of its robotic arm that were then stitched together.

Rovers: An advanced form of landing vehicle that is able to move about on the surface of the world it visits. The first were 'Moon buggies' driven by Apollo astronauts, and robotic versions sent by their Soviet rivals. More recently, this concept has been used several times to survey the surface of Mars.

Penetrators: Some space scientists are planning the design of a new type of impacting spacecraft to tell us more about the composition of other worlds. Flying like javelins, these small missiles would dig themselves into the surface. They could, for example, reveal more about polar ice on the Moon, or the nature of the underground ocean on Jupiter's moon Europa.

FIRST STEPS INTO SPACE

History was made on 4 October 1957 when the Soviet Union became the first nation to put an artificial satellite into orbit about the Earth. Their success, with the launch of Sputnik 1, a beachball-sized orbiter, came after an intense competition with the United States and took their American rivals completely by surprise. US rocket scientists came under pressure to respond quickly – the Space Race had begun.

SHARING THE SPOILS

When Allied forces entered Germany to end World War II in 1945, they found the Nazis' leading rocket engineer Wernher von Braun and hundreds of his scientists and engineers waiting to greet them. Adolf Hitler had ordered they be killed to prevent Germany's rocket secrets being captured, but von Braun saved his team by crossing American lines. Elsewhere in Germany, a Red Army officer and rocket enthusiast called Sergei Korolev gathered V-2 rocket parts and documents and took them back to the Soviet Union along with around 150 German scientists and engineers. Some also went to the UK.

SPUTNIK TAKES THE LEAD

The Soviet team led by Korolev had combined the knowledge they gained from the Germans' V-2 development with missile technology they had been secretly working on themselves. They were actually developing a larger, more complex satellite, but fearing the Americans would beat them into orbit, rushed ahead with the simpler Sputnik 1 instead.

A modified R-7 intercontinental ballistic missile blasted off from a site at Tyuratam, Kazakhstan, that would later become the Baikonur Cosmodrome. Its payload, 58 cm in diameter and weighing 83.6 kg, was put into an elliptical orbit that took it round the Earth every 98 minutes, from an altitude of 215 km to 939 km. Once free from its protective shield, or fairing, four long antennas extended. The satellite carried simple temperature and pressure sensors, plus two radio transmitters. Sputnik's familiar beep was picked up by radio dishes and shocked American citizens watched the spacecraft (in fact they were seeing its main rocket stage) fly across their night sky, shining like a bright star. The satellite re-entered the atmosphere on 4 January 1958, after around 1,400 orbits.

As US rocket scientists raced to catch up, the Soviets followed with a second launch just to rub their noses in it, and this time the spacecraft had a passenger. Laika was a female crossbreed stray dog picked up on the streets

◀ America's Explorer 1 satellite is blasted into space in January 1958 atop a Jupiter C missile renamed Juno for the peaceful mission.

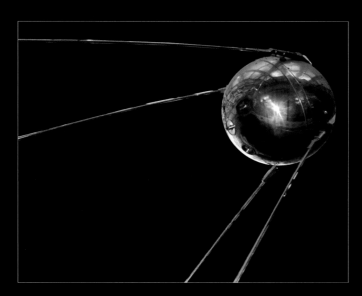

▲ A model of the Soviet Sputnik 1 which became the world's first artificial satellite in October 1957.

of Moscow. She blasted into space on the larger Sputnik 2 on 3 November 1957. Though Laika was given a sealed, temperature-controlled cabin with supplies of air, food and water, it was a one-way trip. Sadly, damage to the probe also meant she must have suffered intense heat and died within a few hours.

THE USA RESPONDS

Somewhat humiliated, the US did not catch up with the Sputnik success until 31 January 1958, with the launch by Wernher von Braun of its first satellite, Explorer 1.

Von Braun and his team had earlier set up shop at White Sands, New Mexico, and made many launches including a two-stage rocket, Bumper 5, that soared to a height of 393 km in February 1949, putting it well above the 100 km high Kármán line marking the boundary of space. A year later, launches switched to a new site at Cape Canaveral, Florida, which would become the USA's primary spaceport.

When Sputnik 1 launched, a rival US Navy team had been testing a spacecraft called Vanguard with successful suborbital flights. They rushed to launch into orbit on 6 December 1957, but the rocket exploded as it lifted off.

Meanwhile, von Braun was separately developing Explorer 1 in secrecy with the US Army Ballistic Missile Agency. It blasted into space from Cape Canaveral, Florida, atop a Jupiter C missile renamed Juno to mark its civilian use. Explorer 1 was just over 200 cm long and weighed

▶ Wernher von Braun, who led rocket development for the United States, stands in front of the Saturn V that flew Apollo 11's crew to the Moon in 1969.

14 kg. It went into a higher orbit than intended, swooping from a distance of 2,515 km to just 354 km every 115 minutes. The satellite made a major scientific discovery by detecting a highly charged zone around the Earth, now known as the Van Allen Belts. It stayed in orbit for 12 years.

The next US launch, of Explorer 2 on 5 March, failed due to a faulty fourth stage. But the rival team finally put their own Vanguard 1 satellite into orbit on 17 March 1958.

▼ A model of Explorer 1's rocket is held aloft by, from left, the Director of NASA's Jet Propulsion Laboratory, William Pickering, scientist James Van Allen and rocket pioneer Wernher von Braun as its successful launch is announced.

THE FINAL FRONTIER

FIRST MAN IN SPACE

Just three and a half years after the launch of Sputnik 1, the Soviet Union achieved an even bigger success by putting the first man in space. Cosmonaut Yuri Gagarin, a pilot in the air force, blasted off on 12 April 1961 to make one orbit of the Earth, at a greater height than intended, in his Vostok spacecraft. Gagarin, who was 27 when he made the historic flight, landed in an unorthodox way. His spherical capsule was not considered safe to land in, so he had to eject when he was about 7 km above the ground and glide the final distance by parachute. He came down about 300 km short of his expected landing spot, so then had to walk to find local villagers to help alert the authorities.

Gagarin's flight made him a national hero and brought him worldwide fame. He never flew in space again, and was tragically killed in a mysterious jet crash in March 1968, aged just 34.

Once again, the United States had to play catch-up, and this they did less than a month after Gagarin's flight on 5 May 1961 when military test pilot Alan B. Shepard flew into space aboard his *Freedom 7* spacecraft, as part of Project Mercury. He was one of seven men in an astronaut corps selected by the National Aeronautics and Space Administration (NASA), which had been created in 1958 to lead the US space programme. Shepard, then 37, might have beaten Gagarin into space had his flight not been repeatedly delayed. His capsule, launched by a Redstone

▲ Cosmonaut Yuri Gagarin, in his spacesuit, on his way to the launch at Baikonur Cosmodrome on 12 April 1961 that will make him the first human in space.

rocket from Cape Canaveral, did not go into orbit, but instead made a 15-minute suborbital flight. It reached a height of 187 km before splashing down in the Atlantic Ocean. With national pride restored, President John F. Kennedy challenged NASA the same month to put a man on the Moon by the end of the decade.

Shepard was followed into space by Virgil 'Gus' Grissom, on 21 July 1961, who nearly drowned when the hatch of his *Liberty Bell 7* capsule blew open on splashdown. But it was not until 20 February 1962 that an American first orbited the Earth. Launched by a Mercury Atlas 6 rocket from Cape Canaveral, John Glenn's *Friendship 7* spacecraft made three orbits, reaching a maximum altitude of about 260 km, before splashing down nearly five hours later in the Atlantic Ocean, to the south-east of Bermuda.

FIRST WOMAN IN SPACE

The Soviets succeeded in sending the first woman into space with the launch of their sixth cosmonaut, Valentina Tereshkova, on 16 June 1963. She piloted Vostok 6 on a mission lasting almost three days, during which she made 48 orbits of the Earth.

NASA did not put a woman into space until June 1983, when Sally Ride flew aboard the Space Shuttle *Challenger*.

◄ The first woman in space, Valentina Tereshkova slips on her gloves during training for her historic flight in 1963.

Astronaut Alan Shepard is picked up by helicopter after splashdown of his *Freedom 7* capsule.

ANIMALS IN SPACE

Apart from the Russian dog Laika, in 1957, the path into space was travelled by several animals before any humans blazed the trail. On the US side they included three monkeys and a mouse that flew on V-2s as early as 1948–9 but died, and others that survived tests in the 1950s. The biggest loss was of a crew of 14 mice on one Jupiter rocket that was destroyed after launch in 1959.

The Soviets were testing with mice, rats and rabbits before settling on dogs, several of which perished in flight. Bitches were chosen over male dogs because there would be no room in their containers to cock their legs.

The earliest 'astronauts' in NASA's Mercury programme were chimpanzees. The first, called Ham, flew a suborbital test before Alan Shepard in January 1961. In November that year, another chimp, Enos, flew into orbit as a dummy run for John Glenn. Both chimpanzees survived their pioneering flights.

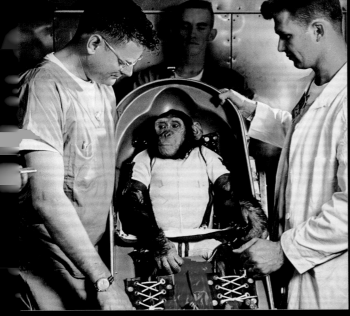

▲ Chimpanzee Ham, who flew a suborbital mission to pave the way for human astronauts.

► Astronaut Virgil 'Gus' Grissom, photographed as he was about to enter *Liberty Bell 7*.

DESTINATION MOON

The Moon's heavily cratered surface is revealed from NASA's Galileo space probe on its way to Jupiter in December 1992. In recent times, only robotic craft have visited our natural satellite, but history was made when 12 men travelled to land on and explore the lunar surface in humanity's greatest adventure. Remarkably the Apollo project achieved its goal only eight years after the first humans flew into space.

THE RACE TO THE MOON

The seeds of the historic Apollo missions were sown just days after America's first astronaut, Alan Shepard, made his suborbital flight in *Freedom 7*, following in Yuri Gagarin's wake to reach space. Having consulted with NASA Administrator James Webb and other experts, President John F. Kennedy was convinced that the United States could regain the initiative from the Soviet Union by aiming for the Moon. He told a special joint session of Congress, on 25 May 1961: 'I believe that this nation should commit itself to achieving the goal, before this decade is out, of landing a man on the Moon and returning him safely to Earth.'

With the Cold War still causing great rivalry between the two superpowers, the US President expanded on his theme the following year in a speech at Rice Stadium, Houston, Texas. On 12 September 1962, in words that have become famous, he declared: 'We choose to go to the Moon. We choose to go to the Moon in this decade and do the other things, not because they are easy, but because they are hard, because that goal will serve to organize and measure the best of our energies and skills, because that challenge is one that we are willing to accept, one we are unwilling to postpone, and one which we intend to win....'

It was a bold ambition. By the time Kennedy spoke in Houston, the US had already made four attempts to send a robotic probe to the Moon with the start of the Ranger programme, designed to help find landing sites. All had failed. Rangers 1 and 2 both failed to leave Earth orbit in 1961, Ranger 3 missed the Moon by 36,000 km with contact lost, and Ranger 4 crashed into the Moon, as intended, but without sending back any images due to a computer fault.

The next two Rangers, 5 and 6, were duds too, the first missing the Moon having lost contact with Earth, and the second crashing but with camera failure. However, there then followed three successful missions to end the series. Ranger 7 impacted Mare Cognitum (Sea of Knowledge) in July 1964, Ranger 8 hit Mare Tranquillitatis (Sea of Tranquillity) in February 1965, and Ranger 9 crashed into crater Alphonsus in March 1965. They sent back photos 1,000 times better than could be taken with telescopes on Earth. Incidentally, these lunar 'seas' are all dry!

Early Soviet efforts to reach the Moon, with the Luna programme, had been largely unsuccessful too, though Luna 3 bucked the trend by flying past and taking the first images of the lunar far side, which had never before been seen by human eyes. Further success came after a flurry

▼ During the robotic missions to the Moon to prepare for Apollo, Lunar Orbiter 1 took the first historic photograph of the Earth above the lunar limb.

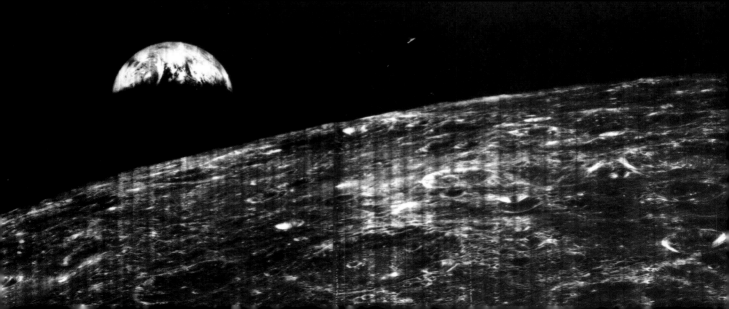

▲ Launch of the rocket carrying Surveyor 1, which made NASA's first soft landing on the Moon's surface.

of launch attempts, when Luna 9 became the first spacecraft to make a soft landing in Oceanus Procellarum (Ocean of Storms), in January 1966. They had once again stolen a march on the United States, which successfully landed its own Surveyor 1 four months later in May, also in Oceanus Procellarum.

The US made four further successful landings in the Surveyor series with Surveyors 3, 5 and 6 in 1967, landing in Oceanus Procellarum, Mare Tranquillitatis, and Sinus Medii (Central Bay) respectively, and Surveyor 7 in January 1968, touching down near the crater Tycho. Surveyors 2 and 4 both crashed. The successful, three-legged Surveyors returned pictures from the Moon's surface, while those from Surveyor 5 onwards also carried out scientific studies of the lunar soil, or regolith. Importantly, they also demonstrated technology that would be developed for Apollo, and showed that, contrary to some fears at the time, landers would not simply sink deep into moondust!

President John F. Kennedy tells the crowd at Rice Stadium, Houston, in September 1962 why he has set the challenge to put a man on the Moon.

MEANWHILE IN ORBIT

As well as carrying out soft landings on the Moon, NASA sent five identical spacecraft to orbit the Moon and photograph it in detail, mainly to locate the smooth, level areas where the first humans could safely land. The Lunar Orbiter missions, launched in August and November 1966, and February, May and August 1967, photographed 99% of the Moon in high resolution, allowing 20 potential landing sites to be narrowed down to eight.

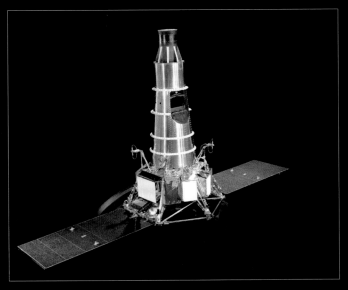

▲ A model of one of the Ranger spacecraft which NASA crashed on the Moon. The aim of the programme was to take close-up images of the lunar surface during their descent.

KENNEDY REMEMBERED

NASA's Florida home, Cape Canaveral, was renamed Cape Kennedy in November 1963, shortly after the President's assassination, but reverted to its old name ten years later. The launch complex is still known as Kennedy Space Center.

WALKING IN SPACE

A landmark in the exploration of space came in 1965 when both Soviet and American astronauts ventured outside their space capsules to make what became known as a spacewalk. Not for the first time, it was a cosmonaut who made the first historic step – and it nearly cost him his life.

Alexei Leonov left the airlock of his two-man Voskhod 2 capsule in March 1965, to begin the EVA (extra-vehicular activity), joined by a 5 m tether to keep him from floating away. For 12 minutes, Leonov took in the incredible view of the Earth and stars. But then he found his spacesuit had ballooned in the vacuum until it was too big to let him back inside. At huge risk of decompression sickness, Leonov managed to release air from the suit until eventually he could squeeze himself back in, head first, to rejoin crewmate Pavel Belyayev.

Three months later, Ed White left the safety of NASA's Gemini 4 spacecraft to carry out his own EVA. Kept secure by an 8 m tether, White used a gun that fired a jet of oxygen to propel him outside, just like a rocket engine. After three times travelling the length of the cord and back again, the gun's fuel ran out and White hauled himself back by hand. America's first spacewalk lasted 23 minutes.

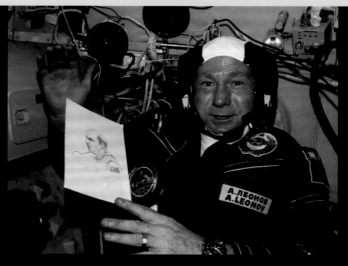

▲ The first spacewalker, cosmonaut Alexei Leonov, is seen here during a later Apollo–Soyuz space mission, waving and holding a sketch he had made of US astronaut Tom Stafford (see page 60).

HEAVENLY TWINS – THE GEMINI MISSIONS

Gemini 4 was one of 12 flights, ten of them crewed, of a new American two-man spacecraft that was developed to get future Apollo crews used to flying together in space for several days – and to prepare them for the experience of a mission to the Moon.

At first glance, the Gemini capsule resembled that used in the Mercury programme, but was larger and more versatile. It could manoeuvre in space to change its orbit, using thrusters.

▼ Astronaut Ed White makes America's first spacewalk on 3 June 1965, tethered to his Gemini 4 space capsule.

The Gemini spacecraft also had to be launched by a bigger rocket, Titan II, an adaptation of an American missile.

The first two Gemini missions were unmanned test flights, but Gemini 3 launched in March 1965 with Gus Grissom and John Young aboard. They splashed down about five hours later in the Atlantic, 111 km short of their target, and both became seasick waiting for rescue.

After the four-day Gemini 4 mission with Ed White and Jim McDivitt, the next crew, Gordon Cooper and Pete Conrad, completed an eight-day flight aboard Gemini 5 in August 1965, despite problems with the oxygen supply and a rocket thruster.

The next two missions – Gemini 7 on a 13-day flight carrying Frank Borman and Jim Lovell, and then Gemini 6 on a day-long trip crewed by Walter Schirra and Tom Stafford – performed the first rendezvous in space, remaining close for more than five hours.

Gemini 8, with Neil Armstrong and David Scott aboard, flew in March 1966, making a historic first docking with an

THREE'S A CROWD

The Soviet Voskhod that carried Alexei Leonov – which was basically a modified Vostok – had made only one other flight, in October 1964, when it became the first spacecraft to carry more than one astronaut. Squeezed inside were three cosmonauts: Vladimir Komarov, Konstantin Feoktistov and Boris Yegorov. The craft was succeeded by the Soyuz, a basic design that would serve for more than 50 years.

unmanned Agena Target Vehicle (ATV). A thruster malfunction caused a near-disastrous tumbling of the docked spacecraft, and the mission was brought to an end after less than 11 hours.

Tom Stafford and Gene Cernan were unable to dock Gemini 9 with their target in June on a three-day flight. But John Young and Michael Collins made a successful rendezvous with the Agena in Gemini 10, in July 1966. Collins spent 49 minutes standing at the open hatch and, later, a further 39 minutes making a spacewalk on the near three-day mission.

Gemini 11, with Pete Conrad and Dick Gordon aboard, docked with the Agena, with Gordon making two excursions outside during the three-day trip. The final Gemini mission was with Jim Lovell and Buzz Aldrin in Gemini 12, in November 1966. They performed a similar docking and Aldrin spent a record 5 hours 30 minutes on a spacewalk. The mission lasted a little under four days.

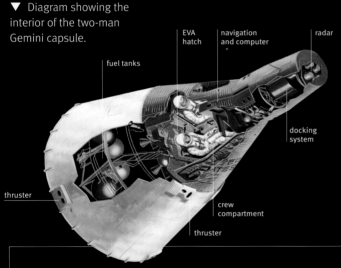

▼ Diagram showing the interior of the two-man Gemini capsule.

EVA hatch

navigation and computer

radar

fuel tanks

docking system

thruster

crew compartment

thruster

▼ The Gemini 7 spacecraft is pictured floating above Earth from Gemini 6, marking the first rendezvous in space.

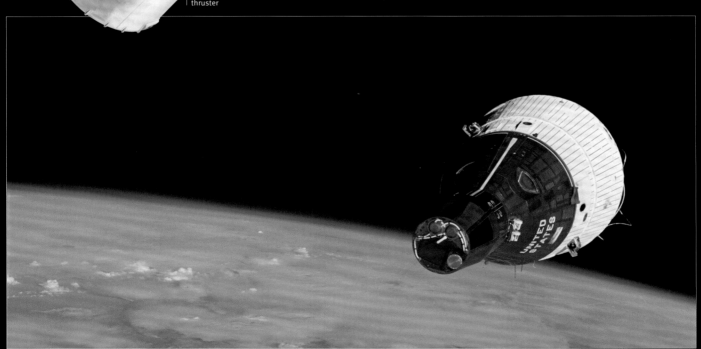

APOLLO – A SPACESHIP FOR THE MOON

To send humans to the Moon, NASA built its biggest rocket ever, a three-stage monster called Saturn V that stood 111 m tall, or about as high as a 36-storey building. It would launch the Apollo missions that made history. Atop the towering rocket was to be a three-part spaceship. One was the Command Module, another conical capsule, but big enough to carry three astronauts in relative comfort. For the trip to the Moon and back, it was attached to a cylindrical Service Module to provide electrical power, propulsion and storage. And tucked behind the assembly, with its three landing legs folded, was the Lunar Module that would ferry two astronauts to the lunar surface and launch them back again.

To reach the Moon, the Saturn V would first put the spacecraft into orbit around the Earth where checks could be made that all was well. Then the third stage would fire its engines again to send it towards the Moon. On the three-day journey there, panels protecting the Lunar Module (LM) would be jettisoned – the Command/Service Module (CSM) would then separate from it and turn to dock with the lander, pulling it away from the rocket stage that is now discarded.

The mated modules then go into orbit around the Moon. After a number of orbits, two members of the crew crawl through a hatch into the Lunar Module; it detaches from the rest of the assembly and fires an engine to descend to the Moon's surface. The third astronaut remains in the orbiting Command Module.

▶ A Saturn V rocket sits on the launch pad at Cape Kennedy, and (inset) a cutaway shows how the crew of three astronauts fitted into their Command Module.

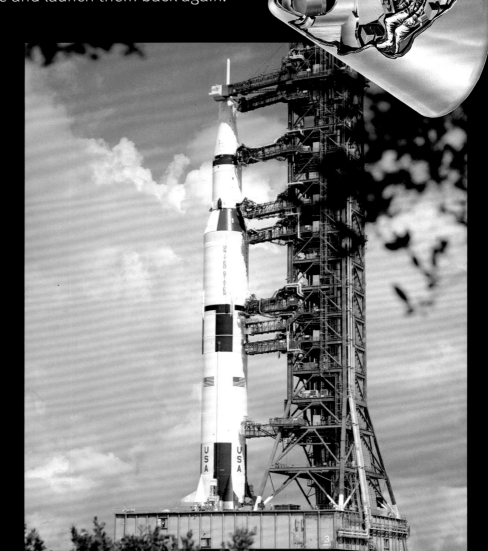

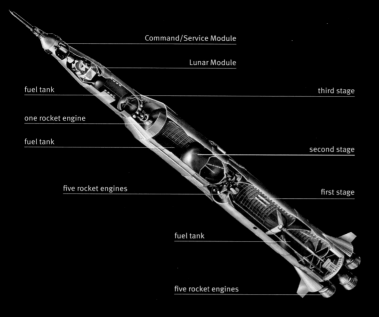

Command/Service Module

Lunar Module

fuel tank   third stage

one rocket engine

fuel tank   second stage

five rocket engines   first stage

fuel tank

five rocket engines

▲ A cutaway diagram shows how the Saturn V rocket was constructed to carry humans to the Moon.

On the Moon, the two astronauts climb down a short ladder from the lander to the lunar surface. Following one or more such excursions, exploring and collecting rock samples, they return to be launched back into space by the Lunar Module's ascent stage. The base of the lander remains on the Moon.

After rendezvousing and docking with the Command Module again, the astronauts rejoin their companion, and the lander is jettisoned before the Service Module's engine fires to send them back towards Earth. Another three-day flight follows, before the Service Module is abandoned and the Command Module turns to descend, heat shield first, for re-entry into the atmosphere. Finally, the capsule parachutes to splash down in the Pacific Ocean to be picked up by an aircraft carrier.

A SMALLER SATURN

The Saturn V had a smaller relation, the Saturn IB, which was used to launch four unmanned test flights of Apollo hardware plus the first crewed mission, Apollo 7, which only went into low-Earth orbit with a light payload.

▶ The Apollo 1 astronauts (from left), Gus Grissom, Ed White and Roger Chaffee, photographed at the launch pad where they later tragically died. Above this image is their mission patch.

FIRE ON THE LAUNCH PAD

Due to the complexity and daring nature of the Apollo missions, it was always feared that there might be a disaster that could cost astronauts their lives. Few imagined that it would happen, not in space, but during a routine test on the ground. Tragically, that is what happened when the crew for the first planned manned mission, Apollo 1, were carrying out a pre-flight exercise on the launch pad at Cape Kennedy, Florida, in January 1967.

Gemini veterans Gus Grissom and spacewalker Ed White, together with rookie astronaut Roger Chaffee, were trapped inside when an electrical fault caused fire to break out inside the Command Module. Fed by pure oxygen, the blaze raged as the astronauts found themselves unable to open the hatch. The intense heat ruptured the capsule allowing air to flood in, smothering the flames but creating dense smoke that killed the three men.

An inquiry into the disaster led to a redesign of the hatch so that it opened outwards, and a change in air supply to a mix of oxygen and nitrogen rather than pure oxygen. Flammable materials in the cabin were replaced with fire-resistant ones. The first crewed Apollo mission, which had been due to fly in February 1967, did not launch until October 1968.

APOLLO 7: ASTRONAUTS GET THE TRAVEL BUG

The first Apollo mission to carry a human crew was numbered Apollo 7. The first US spaceflight to fly with three astronauts, it was also probably the grumpiest mission ever after all three went down with head colds. The mission, launched by a Saturn IB rocket in October 1968, was planned to last 11 days and provide a test of the combined Command and Service Modules (CSM) in low-Earth orbit. No Lunar Module was carried.

Trouble began soon after launch when the crew's commander, space veteran Walter Schirra, reported symptoms of a cold developing. A day later, the two other astronauts, Donn Eisele and Walter Cunningham, started to go down with the bug too.

The crew's symptoms were made worse in zero gravity because fluids could not drain from the head. They found it so uncomfortable that they rebelled against some commands from mission control at Houston and finally refused to wear their space helmets during re-entry, claiming it would prevent them relieving pressure on their sinuses.

The mission, which was the first crewed trip to broadcast live on television from space, ended when the crew's capsule splashed down in the Atlantic Ocean. But after their mini mutiny, none of the crew ever flew in space again.

▲▶ The Apollo 7 crew (above, from left), Donn Eisele, Walter Schirra and Walter Cunningham are all smiles as they pose before their mission, but the strain later shows on the face of Commander Wally Schirra (right), who was suffering the effects of a head cold.

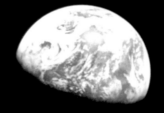

APOLLO 8: CHRISTMAS IN SPACE

Two months after Apollo 7, the next mission in the programme blasted into space, this time hoisted by a giant Saturn V rocket. Apollo 8, with its crew Commander Frank Borman, Jim Lovell and Bill Anders, set new space records as it became the first manned spacecraft to leave Earth's orbit and fly to the Moon.

The Apollo 8 mission was an important dress rehearsal for the Moon landings to come. The difference was that it did not carry a Lunar Module, just a dummy as ballast to balance the weight within the spacecraft assembly. A six-day adventure, the spacecraft's launch came on 21 December 1968, meaning that the crew would be away over Christmas. They marked the holiday by reading from the Bible's Book of Genesis on Christmas Eve in a live TV broadcast from their Command Module, now orbiting the Moon. They ended by wishing everyone on Earth a 'Merry Christmas'. It was one of six broadcasts made.

As well as testing the spacecraft's communications, hardware and life-support systems, and seeing how astronauts responded on a flight far from home, the mission allowed the crew to photograph much of the lunar surface. They took many photographs of the near and far sides of the Moon, gathering information that would help with the upcoming landings.

RUSSIAN RIVALRY

After the United States announced its goal to reach the Moon by the end of the 1960s, Soviet space chiefs began work in secret to make their own bid to get cosmonauts there first. Separate missions were independently designed, based on their new Soyuz spaceship design: one to fly around the Moon and back, and the other to orbit the Moon and land in a similar fashion to Apollo. The giant N1 rocket that was being developed to carry a two-man crew to the Moon suffered several failures, including a catastrophic explosion on one test launch. After the success of Apollo 8 in reaching the Moon, the Soviets realized they had lost the race and their secret programmes were quietly wound down.

The astronauts also became the first humans to see the Earth rising above the lunar horizon, a spectacle that provided one of the mission's most memorable images.

After ten orbits of the Moon, the Service Module's burn on Christmas Day sent the astronauts heading back home again. They splashed down in the Pacific Ocean, 1,600 km south-west of Hawaii, to be picked up by the USS *Yorktown*.

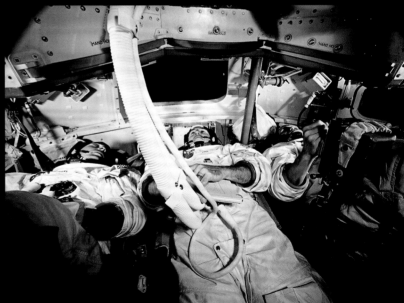

▲ The Apollo 8 astronauts (from left), Anders, Lovell and Borman, in training for the mission that would take them around the Moon.

◄ The roar of the Saturn V launching Apollo 8 to the Moon startles the birds at the Kennedy Space Center, Florida.

◄◄ (Far left) A dramatic view of the cloud-shrouded Earth rising over the lunar landscape, photographed by the crew of Apollo 8.

APOLLO 9: SPIDER MAKES FIRST FLIGHT

The next flight to help prepare for a Moon landing was also the first crewed mission to see the Lunar Module fly, but it was a flight that did not leave Earth orbit. Apollo 9, which launched in March 1969, was designed to test out all the manoeuvres that the space vehicles would make on the Moon trip, but at a relatively safe distance.

The mission was commanded by Jim McDivitt, with David Scott and Rusty Schweickart as crew. After launch by a Saturn V, the combined Command/Service Module (CSM), nicknamed *Gumdrop*, separated from the rocket's third stage, which still had the Lunar Module (LM), dubbed *Spider*, attached. Panels protecting the LM were jettisoned and, three hours after launch, *Gumdrop* turned to dock with *Spider*.

The astronauts rehearsed rescue operations in case the LM lost power, firing *Gumdrop*'s engines to steer it, and also by carrying out spacewalks, or EVAs, from both the CSM and the LM while they were still attached.

Then the LM, piloted by Schweickart and with McDivitt aboard, separated from the CSM, with Scott at the controls, to fly as an independent spacecraft and simulate moves it would make on a Moon mission. The LM's descent stage was jettisoned as the ascent stage fired its engine, simulating the launch that would return astronauts from the Moon's surface. *Spider* was now more than 120 km behind *Gumdrop*, but it returned to rendezvous with the mother ship, just as its successors would from the lunar surface.

Once McDivitt and Schweickart were back in the Command Module, the rest of the space hardware was abandoned and the crew splashed down in the Atlantic Ocean with the mission declared a success. The mission also produced some of the most spectacular imagery from an Apollo mission, with the Earth a highly photogenic backdrop.

▲ Astronaut David Scott stands in the open hatch of the Lunar Module *Spider*, docked with the Command Module *Gumdrop*, during the Apollo 9 test flight around the Earth.

▶ The Lunar Module *Spider* has its landing gear deployed above an ocean backdrop during test manoeuvres on the fifth day of the Apollo 9 mission in Earth orbit.

APOLLO 10: PEANUTS IN SPACE

In May 1969, another Saturn V rocket launched the next mission in the series, Apollo 10. This was the final dry run for a historic landing. It flew to the Moon and completed all elements of what would be required, apart from touching down on the lunar surface.

The crew were Commander Tom Stafford, John Young and Gene Cernan, all veterans of the Gemini programme. The CSM, piloted by Young, was dubbed *Charlie Brown* after the character in the famous *Peanuts* cartoon strip, and the LM was nicknamed *Snoopy* after his dog.

After a three-day trip to the Moon, during which the two spacecraft modules docked, Apollo 10 went into a circular orbit around the Moon. Stafford and Cernan then boarded the LM and fired its thrusters to separate the two craft.

Snoopy changed its orbit to allow the module to swoop low over the lunar surface, coming to within 14 km from the ground. This allowed the astronauts to take many

photographs, particularly of the sites picked for the future landings. Following extensive tests, *Snoopy*'s descent stage was jettisoned to crash eventually into the Moon, and the ascent stage rendezvoused and docked again with *Charlie Brown*, eight hours after they had parted company.

The ascent stage was then discarded, into an orbit that sent it circling the Sun, and the next day *Charlie Brown* fired its engines to head for home, having made 31 orbits of the Moon. Two days later, the Command Module broke free from the Service Module and delivered the crew safely to a splashdown in the Pacific.

Apollo 10's Command Module is the only one that now has a home outside the United States after it was presented on extended loan for display at the Science Museum in London, England.

▶ A photograph of the Apollo 10 crew – seen from left to right are Commander Gene Cernan, Tom Stafford and John Young.

▼ A view from the Lunar Module *Snoopy* of the Command Module *Charlie Brown* after separation, while orbiting the Moon.

MAN ON THE MOON

In July 1969 came the moment towards which America's space programme had been working since President Kennedy set the challenge in 1961 – humans set foot on the surface of the Moon. The humans in question were Neil Armstrong and Edwin 'Buzz' Aldrin. Armstrong made the historic first step, as commander of the Apollo 11 mission. He did so with some less than spontaneous first words, declaring: 'That's one small step for a man; one giant leap for mankind.'

The mission had begun on 16 July when the astronauts' spacecraft was hoisted aloft by a Saturn V rocket from the Kennedy Space Center, Florida. Together with Armstrong and Aldrin was the third member of their crew, Michael Collins. All three had previously flown on Gemini missions.

As with previous Apollo missions, the combined spacecraft first went into Earth orbit, for one and a half orbits, before the rocket's third stage burned its engines again to head for the Moon. As with Apollo 10, protective panels around the Lunar Module, which was named *Eagle*, were ejected and the Command/Service Module, called *Columbia*, turned to dock with it and pull it free.

▼ History is in the making as Buzz Aldrin stands on the surface of the Moon, his footprints clearly visible in the lunar soil. Photographer Neil Armstrong and the Lunar Module *Eagle* can be seen reflected on his visor.

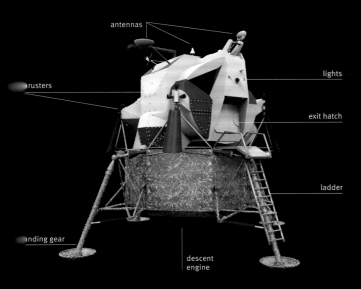

antennas

lights

thrusters

exit hatch

ladder

landing gear

descent engine

▲ A model of the Lunar Module that carried the first astronauts to the surface of the Moon and launched them home again.

Two colour TV broadcasts were made from the spacecraft on the outward voyage, the first from *Columbia*, and the second from *Eagle* after Armstrong and Aldrin put on their spacesuits and crawled through to examine the lander.

On 19 July, while Apollo 11 was behind the Moon and out of contact, the main engine was fired to put the spacecraft into lunar orbit. It was a tense moment as mission controllers and the rest of the world watching history in the making waited for contact to be resumed to find out if it had been successful.

The following day, Armstrong and Aldrin entered the Lunar Module again to check it out once more before the lander separated from *Columbia*, extending its folded legs. Collins remained in the mother ship as it continued to orbit the Moon. On its 13th orbit, and on the far side of the Moon, *Eagle* fired its engine to begin descent. The challenge was to bring it from a height of 15,000 m, orbiting at several thousand kilometres per hour, to a soft touchdown.

A second, longer firing, after the Lunar Module had re-emerged from behind the Moon, brought the lander lower until it was about 8,000 m from its intended landing site and 8 km above the lunar surface.

Landing *Eagle* was always going to be a challenge for Armstrong and Aldrin, and they almost ran out of fuel before they made it to the surface. The trouble began when they were close to their planned landing spot in Mare Tranquillitatis (Sea of Tranquillity) and warning lights began flashing as the onboard computer struggled to keep up with commands. Then, as Armstrong was looking out of the window for a smooth area to put *Eagle* safely down, he was alarmed to see that they were heading directly for a large crater in an area strewn with car-sized boulders. Attempting a landing there risked tipping the spacecraft over, which would have been a disaster.

Armstrong quickly took over control from the Lunar Module's computer and steered it over the crater and boulders to search for a safer landing zone. As he finally started to bring the craft down, and the thruster began disturbing the moondust, mission control at Houston warned that they were close to having to abort the landing. But then the lander's legs gently touched the surface and the craft was safely down. It was 20 July.

Armstrong radioed home to announce: 'Houston, Tranquillity Base here. The *Eagle* has landed.' There was just 20 seconds of fuel left in the tank. Fellow astronaut Charlie Duke, who was channelling communications from mission control, replied: 'Roger, Tranquillity. We copy you on the ground. You got a bunch of guys about to turn blue. We're breathing again. Thanks a lot.'

▲▶ The crew of Apollo 11 (above) – seen from left to right are Commander Neil Armstrong, Michael Collins and Buzz Aldrin. (Right) The three men blast off into space atop their Saturn V rocket.

APOLLO 11: ONE SMALL STEP

With the *Eagle* lander safely on the Moon, the watching world was impatient to see the historic moment when Neil Armstrong and Edwin 'Buzz' Aldrin would become the first humans to walk on the lunar surface. The astronauts were equally keen to do so. They were due to get four or five hours of rest first, having been awake for many hours, but understandably they believed they would have trouble sleeping at such an exciting time.

Mission control allowed them to commence preparations to leave the Lunar Module for their EVA. Even so, it was more than six hours before they would step on to the Moon's surface. Armstrong was first to exit the hatch and descend the ladder. A camera attached to the craft sent back ghostly black-and-white images of his progress to be televised live around the Earth.

Finally, at 02:56 Universal Time on 21 July (when it was still the evening of 20 July in the United States) Armstrong stepped off the ladder and on to the Moon, to say the now famous words that he had planned on the journey there.

After Armstrong checked the surroundings and collected some samples of dust and rock in case they needed to make a quick getaway, Aldrin descended the ladder to join him on the surface. The two men, weighing a sixth of what they would on Earth, thanks to the Moon's weaker gravity, paused to take in the view that no humans had ever enjoyed before. Aldrin later described it as 'magnificent desolation'.

▶ Buzz Aldrin, the second man on the Moon, pauses beside the flag of the United States (below), and retrieves experiments from the Lunar Module *Eagle* (right).

Their first task was to plant a camera 10 m from the lander, along with a US flag. During more than two and a half hours on the Moon, Armstrong and Aldrin set up a number of experiments to detect moonquakes, collect data on the solar wind and moondust, and allow precise measurements of the Moon's distance from Earth. They also received a phone call from President Richard Nixon, who told them: 'This certainly has to be the most historic telephone call ever made from the White House.'

Their final action – which they almost forgot – was to leave a pouch containing an Apollo 1 patch plus medals to commemorate the astronauts, American and Soviet, who had lost their lives in the quest to reach space. It also held a small silicon disk with messages of goodwill from 73 nations around the world, and a gold olive branch, the symbol of peace.

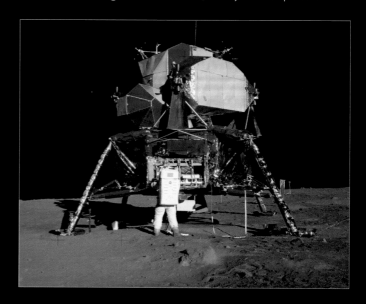

▲▶   A footprint left by one of Aldrin's boots in the lunar soil
will stay undisturbed for eternity. The backdrop (above right)
is an image of the Full Moon, taken from Apollo 11.

Then, after 21 hours 36 minutes on the lunar surface,
including a period of sleep at last, it was time to leave.
The *Eagle*'s ascent stage engine fired to send them back into
orbit. Left behind, attached to the descent stage, a plaque
read: 'Here men from the planet Earth first set foot upon the
Moon July 1969, A.D. We came in peace for all mankind.'
    Michael Collins, who had been patiently waiting in the
orbiting *Columbia*, filmed the ascent stage's approach before
his crewmates docked and crawled back into the Command
Service Module. With no further use for the *Eagle*, it was
jettisoned, before the CSM fired its engine to return to Earth.
    Two and a half days later, the Command Module splashed
down in the Pacific where the USS *Hornet*, with President
Nixon aboard, was waiting to recover the returning heroes.
All three went straight into quarantine for three weeks in case
they had brought back any Moon bugs.

FIRST MAN ON THE MOON

Neil Armstrong was a modest man who kept a low profile after
his historic achievement of becoming the first human to set
foot on another world. He had made a name for himself in
the Korean War in the early 1950s, flying several combat
missions. He went on to become a test pilot. His cool manner
first showed when all four engines of a B-29 Superfortress he
was flying failed. The plane landed safely.
    Then, as a Gemini 8 astronaut he managed a drama when
his capsule went into a spin. Later, he was training in a test
model of the lunar lander, dubbed a 'flying bedstead', when
it went out of control and he had to eject. Coolly, he went
straight back to work in his office.
    After Apollo, Armstrong became a university lecturer,
teaching aerospace engineering. He died in 2012 aged 82.

MOON LANDING FACT FILE

- The *Eagle*'s shock absorbers did not compress on landing, so Armstrong's 'one small step' to the surface was actually a 1 m leap from the ladder.
- A switch needed to fire the *Eagle*'s engine to leave the Moon was broken so Aldrin made a makeshift replacement with his pen.
- The US flag got blown away by the blast as the Lunar Module's ascent stage took off from the Moon.
- Arriving in Hawaii, the Apollo 11 crew had to fill out a customs form to declare where they had just come from.

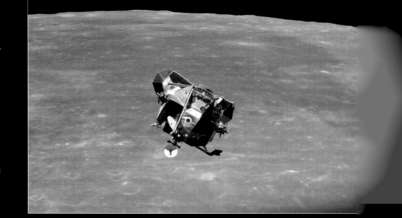

▲   The *Eagle*'s ascent stage is pictured by Command Module
pilot Michael Collins as it returns to join the mother ship,
Columbia.

APOLLO 12: A LIGHTNING RETURN TO THE MOON

America's second crewed mission to the lunar surface got off to a dramatic start when the Saturn V rocket carrying Apollo 12 was twice hit by lightning as it ascended into the sky. The launch, from Kennedy Space Center, Florida, on 14 November 1969, was being watched by President Richard Nixon and went ahead despite local cloud and rain. On board were mission Commander Pete Conrad and Alan Bean, who would walk on the Moon, plus Dick Gordon who would stay with the Command/Service Module, named *Yankee Clipper*. Bean was making his first spaceflight.

The two lightning strikes briefly cut off power in the spacecraft and contact with mission control, but the rocket's first stage continued to fire, lifting the crew towards Earth orbit. The spacecraft momentarily switched to battery mode before the astronauts restored main power and checked out the electrical system.

▼ The view from *Yankee Clipper* as the Lunar Module *Intrepid* descends, carrying Pete Conrad and Alan Bean, to land on the Moon.

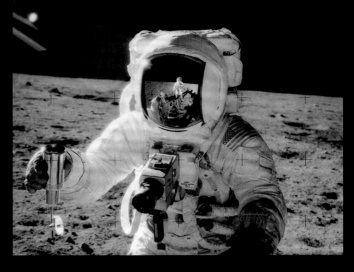

▲ Astronaut Alan Bean is pictured with a container that is holding a sample of lunar soil. Photographer Pete Conrad can be seen in the reflection on his visor.

Once *en route* to the Moon, the routine procedure of docking with the Lunar Module, dubbed *Intrepid*, was performed before Conrad and Bean entered the lander to check for any possible storm damage. Fortunately, none was found.

The spacecraft was put into lunar orbit by a rocket firing while on the far side of the Moon. A second burn two orbits later set things up for the landing. Conrad and Bean entered *Intrepid* and on 19 November the Lunar Module separated to begin its descent.

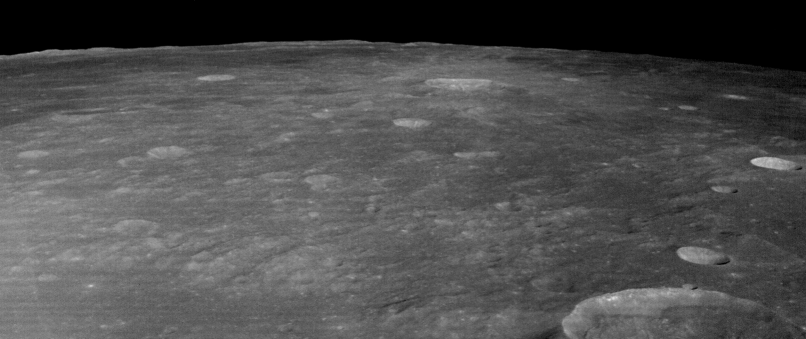

Their target was a landing site in Oceanus Procellarum (Sea of Storms), and with Apollo 11 having landed several kilometres from its intended spot, NASA was keen to make *Intrepid*'s touchdown as precise as possible. It was a much dustier location, and the cloud kicked up by the descent thruster obscured Conrad's view in the final moments before landing. But *Intrepid* came down bang on target and just 164 m from NASA's robotic lander Surveyor 3, which had been sitting there since April 1967.

A few hours later, the first EVA began as Conrad stepped down the ladder and on to the lunar surface. And unencumbered by the demands of making history, his first words were a lot less formal than Neil Armstrong's had been four months earlier. Referring to the Apollo 11 commander's 'One small step', Conrad declared: 'Whoopee! Man, that may have been a small one for Neil, but that's a long one for me.'

The first excursion by Conrad and Bean, lasting just under four hours, was spent setting up a more complex set of instruments on the Moon than the previous mission, known as the Apollo Lunar Surface Experiments Package, or ALSEP. One failure came when Bean was setting up the TV camera and it was accidentally pointed at the Sun, destroying its sensitive electronics. However, the moonwalkers' hand-held camera meant they brought many photographs of their adventure back to Earth.

On 20 November, the pair began a second EVA, collecting a useful batch of rock samples from the ground and below the surface for scientists to study. But the highlight of the

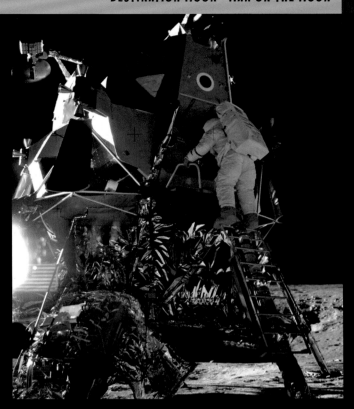

▲ Alan Bean begins his climb down *Intrepid*'s ladder to join Pete Conrad on the lunar surface in Oceanus Procellarum.

moonwalk that lasted 3 hours 49 minutes was when they visited Surveyor 3 and removed parts of it to take home. On their return, it was found that bacteria taken to the Moon by the probe had survived the hostile conditions.

Intrepid's ascent stage lifted off the Moon's surface 31 hours 31 minutes after landing to dock with *Yankee Clipper* and reunite the three crew members. The Lunar Module was then jettisoned to crash deliberately on the Moon, producing an artificial 'moonquake' that the seismology experiment left behind could learn from.

The journey home ended when the Command Module separated from the Service Module and splashed down in the Pacific, where the astronauts were picked up by the USS *Hornet* on 24 November.

▲ The rare sight of two lunar landers in one photograph, as Pete Conrad inspects robotic probe Surveyor 3 with *Intrepid* in the background.

▶ The Apollo 12 crew (from left to right) of Pete Conrad, Dick Gordon and Alan Bean, during training.

APOLLO 13: HOUSTON, WE'VE HAD A PROBLEM HERE

Lightning strikes might have added a touch of drama to the Apollo 12 mission, but it was nothing compared to the challenging full-scale emergency that faced Apollo 13. An explosion on the way to the Moon, more than 300,000 km from home, raised fears that its crew would be lost in space and never return. But the threat of disaster was turned into one of the greatest triumphs in space exploration history when NASA pulled off a seemingly impossible rescue. All plans for a landing on the Moon had to be abandoned, but ironically it was the Lunar Module, *Aquarius*, that saved the stricken crew.

The mission had started well enough for veteran astronaut Commander Jim Lovell, and his crew Fred Haise and Jack Swigert who were on their first spaceflights. One engine on their Saturn V rocket cut out too soon during launch on 11 April 1970, leaving four others to burn longer to put them into Earth orbit. But now they were safely on their way to the Moon and a visit to the Fra Mauro highlands.

After two days, the crew had just finished a lengthy TV broadcast home when Swigert flicked a switch to stir oxygen tanks in the Service Module. A short circuit sparked a fire that caused an explosion in one tank, damaging another and blowing the side off the spacecraft. Electricity, light and water were lost.

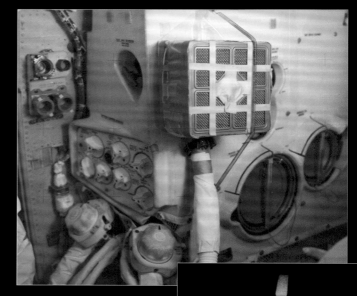

▲ The contraption that the Apollo 13 astronauts had to build from cardboard, plastic bags and tape in order to remove carbon dioxide from *Aquarius*.

▶ After the Service Module was jettisoned, the crew photographed the blast damage from their 'lifeboat'.

Swigert called mission control to report the incident, telling them the now famous words: 'Houston, we've had a problem here.' Warning lights showed that one oxygen tank was empty and the other emptying fast. Glancing out of the window, Lovell could see the precious gas venting into space.

It was hugely fortunate that the Command Module, named *Odyssey*, had already docked with Lunar Module *Aquarius*, because it would have been impossible to carry out such a manoeuvre with a dead spacecraft. As it was, the crew was going to have to rely on the lander to get them home, performing as a lifeboat in a way that it had never been intended to do. *Aquarius* still had oxygen and water in its tanks, plus power and a functioning rocket engine. But it was impossible simply to turn the spacecraft round to bring it home. Instead *Aquarius* would be used to swing the astronauts around the back of the Moon and towards Earth again. The crew would have to navigate, like sailors of old, by aligning a sextant on the Sun.

◀ Apollo 13 lifts off from Cape Kennedy, Florida, on its fateful journey that became a great story of space survival.

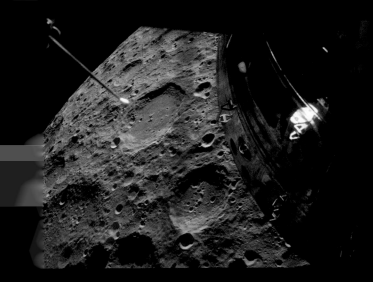

◀ So near and yet so far. A view of the Moon from Apollo 13's Lunar Module *Aquarius* as the crew used it as a lifeboat. The Command Module *Odyssey* is visible at the right-hand side of the image.

was designed to handle and it was building to a perilous level Square canisters in *Odyssey* were designed to remove carbon dioxide, but they were not compatible with the round openings in *Aquarius*'s environmental system. NASA's engineers had to find a way to adapt cardboard, plastic bags and tape on the spacecraft to make the connection.

Before re-entry, the astronauts had to return to the Command Module. They found it cold with condensation covering the walls. They jettisoned the Service Module, taking photos to show the extensive damage from the explosion, then prepared for re-entry. Back home, the world held its breath, but re-entry was successful and on 17 April the astronauts were recovered by the USS *Iwo Jima*.

The astronauts were ordered into *Aquarius* and *Odyssey* was completely powered down. The lander's thruster was fired to send them around the Moon, but calculations showed they would splash down in the Indian Ocean where there would be no rescue ships. A second engine burn was therefore made, both to bring them back more quickly, and to land them in the Pacific. For Lovell, who had been on Apollo 8, it was the second time he had rounded the Moon without landing.

Power use in *Aquarius* was kept to a minimum to conserve it for the long voyage ahead, bringing the temperature down to a chilly 3°C for the uncomfortable crew, who were losing weight and becoming dehydrated.

Then another danger became apparent: the astronauts were producing more carbon dioxide than the Lunar Module

▶▼ (Right) The Command Module *Odyssey* is lifted on to the deck of the rescue ship. (Below) Safely home, the Apollo 13 crew (from left to right) of Fred Haise, Jim Lovell and Jack Swigert.

APOLLO 14: GOLF ON THE MOON

The next NASA mission to the Moon did not fly until January 1971 after modifications were made to the Service Module to avoid an accident such as had hit Apollo 13 from happening again. Apollo 14 lifted off from Kennedy Space Center on 31 January, its destination the Fra Mauro hills that had been the intended landing site for the previous abandoned landing.

The mission was commanded by Alan Shepard, who had been America's first man in space, on his second spaceflight. With him were Stuart Roosa, the Command Module pilot, and Ed Mitchell, as Lunar Module pilot. Both were rookie astronauts making their one and only trips into space.

As they left Earth orbit for the Moon, the astronauts had some trouble docking with the Lunar Module, *Antares*. Six attempts had to be made before the catches successfully connected. After that, it was plain sailing for the rest of the outward voyage, and Shepard and Mitchell landed on 5 February, leaving Roosa to orbit in the Command/Service Module, *Kitty Hawk*.

The two men made two excursions from their lander, spending a total of 9 hours 23 minutes outdoors. They set out the usual ALSEP experiments and also collected more than 42 kg of lunar rock samples that would be shared with scientists around the world. At the end of the second EVA, Shepard used an improvised golf club to hit two golf balls that he had brought with him.

▲ Apollo 14's Command Module splashes down in the Pacific Ocean, ending the three astronauts' ten-day mission.

Antares' ascent stage blasted off from the Moon on 6 February to rejoin *Kitty Hawk* in orbit. The Lunar Module was then sent crashing into the Moon, to provide more data for the seismology experiments, before the mother ship headed home. The astronauts splashed down in the Pacific Ocean on 9 February, to be picked up by the USS *New Orleans*.

▼ Apollo 14 Commander Alan Shepard stands by the US flag in the lunar highlands during the first venture outside.

APOLLO 15: THE CREW BROUGHT WHEELS

With Apollo 14 having put the programme back on track, NASA decided to make the remaining Moon missions more ambitious, giving their crews longer on the lunar surface and providing transport in the form of the Lunar Roving Vehicle (LRV), or 'Moon buggy'.

Commanded by veteran astronaut David Scott, Apollo 15 blasted off from Florida on 26 July 1971, with first-timers Al Worden and Jim Irwin as pilots of the Command Module (*Endeavour*) and Lunar Module (*Falcon*), respectively.

Once in lunar orbit, there was a slight delay when *Falcon*, with Scott and Irwin aboard, failed to undock, but then Worden found a plug had come loose and reconnected it, allowing the manoeuvre to take place. They landed on 30 July in the Mare Imbrium (Sea of Showers) at the foot of the Apennine Mountains, close to a winding canyon called Hadley Rille.

Scott and Irwin spent 18 hours 37 minutes exploring their surroundings during three EVAs, including peering into the canyon, and drove more than 28 km in their buggy. After setting up the ALSEP instruments, they collected more than 77 kg of rock and soil to take home, including a sample from 3 m beneath the surface, and a stone dubbed the Genesis rock because it is nearly as old as the Solar System. For the benefit of TV viewers, Scott dropped a hammer and feather at the same time to show how both would fall at the same rate in a vacuum, just as Galileo had suggested centuries before. He also franked a postage stamp.

After leaving a small sculpture on the lunar surface to commemorate fallen astronauts, Scott and Irwin lifted off and returned to *Endeavour*, before the lander was crashed into the Moon. They also launched a small satellite from the CSM into orbit around the Moon.

► The Lunar Roving Vehicle which allowed Apollo 15's crew to explore more of their surroundings at the Hadley–Apennine landing site. In this photo, Jim Irwin is busy at the rover.

On the way back to Earth, Worden made the first ever spacewalk in deep space to collect film cassettes from mapping cameras at the rear of the Service Module.

Despite the failure of one of its three main parachutes, *Endeavour* splashed down safely, on 7 August, in the Pacific where the USS *Okinawa* was waiting to greet the crew.

▼ A diagram of the Lunar Roving Vehicle. or 'Moon buggy', used by the astronauts on their EVAs.

high-gain antenna
low-gain antenna
film camera
rear storage
hand controller
TV camera
communications unit
dust guards
wire-mesh wheels

APOLLO 16: ROVER DRIVING IN TOP GEAR

A member of Apollo 10's crew, John Young, took command of NASA's next moonshot, Apollo 16, which launched from Florida on 16 April 1972. His crew were Ken Mattingly, pilot of Command Module *Casper*, and Charlie Duke, in charge of Lunar Module *Orion*. They were sent to investigate the lunar highlands in the Descartes region.

The crew encountered a couple of annoying snags with the Command/Service Module, particularly in lunar orbit where *Casper*'s propulsion system was found to oscillate after *Orion* had undocked. The problem threatened to abort the Moon landing, but after a patient wait of nearly six hours, Young and Duke were finally given the go-ahead to descend.

They touched down on the Descartes plain on 20 April where they unloaded their Moon buggy and set up an ultraviolet camera to image the Earth and astronomical targets in the sky. Lunar experiments were set up, but one to measure heat flow in the Moon was destroyed when Young tripped on its cable. Sensors were planted in the lunar soil, or regolith, and explosive charges fired to learn more about the subsurface.

Young and Duke then boarded the rover to drive and inspect two craters and collect samples. A portable device also measured the strength of the Moon's magnetic field. Some hard driving was done to put the rover through its paces on rough terrain and see how well it performed. In all, the astronauts drove 4 km during an EVA lasting 7 hours 11 minutes.

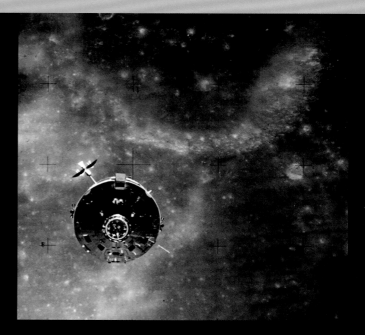

▲ A view of the Apollo 16 Command/Service Module from the Lunar Module *Orion* while in orbit over the lunar surface. This photograph was taken during the 12th revolution around the Moon.

After resting back in *Orion*, the two men began their second excursion by heading south to inspect a crater a little way uphill. Six stops were made to collect rock samples. After 7 hours 23 minutes and 11 km of driving, the astronauts returned to the Lunar Module.

The third and final EVA saw the astronauts visit a large crater with steep walls, and two big boulders during an 11.5 km drive. They also collected samples and retrieved film and data from surface experiments.

Lift-off of the ascent stage on 24 April was broadcast by a TV camera left on the Moon. *Orion* docked with *Casper*, and on the flight home, Mattingly made a spacewalk to collect film canisters while Duke stood in the hatch.

Casper splashed down in the Pacific on 27 April, close to the recovery ship USS *Ticonderoga*.

▼ Charlie Duke collects Moon samples at the rim of Plum Crater, with the parked lunar rover visible in the background.

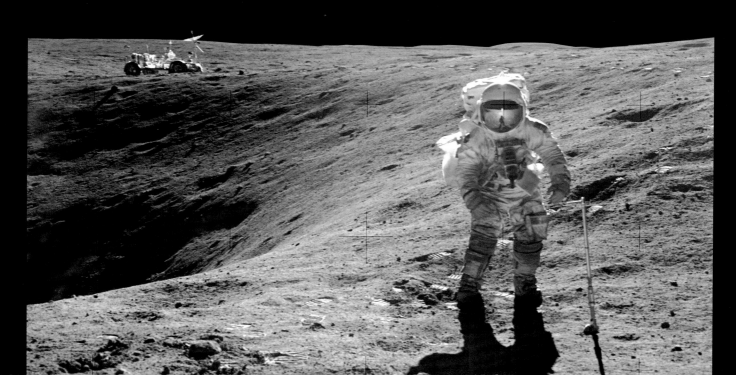

APOLLO 17: APOLLO'S GOODBYE TO THE MOON

Originally, ten lunar landings were planned for Apollo crews. But with Apollos 18 to 20 cancelled in 1970, the final visit to the Moon by humans in the 20th century was aboard Apollo 17, in December 1972. The mission was commanded by veteran astronaut Gene Cernan, with Ron Evans piloting the Command Module (*America*) and professional geologist Harrison Schmitt in charge of the Lunar Module (*Challenger*).

Apollo 17 was NASA's first night launch in the programme, lifting off 2 hours 40 minutes late from the Kennedy Space Center on 7 December. After a routine flight, Cernan and Schmitt descended in *Challenger* to a site on the south-eastern rim of Mare Serenitatis (Sea of Serenity), in the Taurus–Littrow highlands. There they hoped to find the youngest rocks yet, to be picked out by Schmitt's expert eye.

With the rover unloaded and experiments deployed, the astronauts began to explore the area. In three excursions

▲ Apollo 17 Commander Gene Cernan makes a test run in the lunar rover during the first excursion after landing.

they would drive 30 km and collect 110.52 kg of lunar rocks. A highlight was the discovery of orange soil near Shorty Crater, a colourful sight among the otherwise grey surroundings.

With the third excursion completed, the pair returned a final time to *Challenger*. They blasted off to return to the Command Module *America* and headed home, with Evans spacewalking to collect film on the way. They landed safely in the Pacific to be picked up by the USS *Ticonderoga* on 19 December.

Gene Cernan would be the last man on the Moon for several decades, but his footprints, like those of 11 other Apollo astronauts, will remain in the lunar dust for an eternity.

◄ Geologist-astronaut Harrison Schmitt collects samples of rock on a crater rim at the Taurus–Littrow landing site.

▼ Harrison Schmitt stands beside a huge lunar boulder with the lunar rover visible beyond.

DISCOVERY AND DETENTE

America's success in putting men on the Moon was a great achievement in itself. But NASA also took the opportunity to carry out scientific research that would have been impossible otherwise. Results from the experiments, combined with the treasure trove of information from lunar rocks brought back to Earth, helped boost our understanding of the Moon. Later, in 1975, a final Apollo flight was made to perform a symbolic co-operative mission with the Soviet Union.

APOLLO'S LABORATORY ON THE MOON

The Apollo Moon missions carried a variety of instruments to be placed on the Moon's surface in order to tell us more about our natural satellite and its environment. They were known as the Apollo Lunar Surface Experiments Package (ALSEP), though Apollo 11 carried a simpler selection.

The experiments were designed to answer questions not only about the lunar surface, but also about what lay deep within the Moon, how it was structured, and its age. Though the experiments were given an operating life of between one and two years, they continued to send data

home for up to eight years until NASA switched them off. They were stored together in the Lunar Module on a carry bar, to help astronauts unload them, so that they resembled an unwieldy pair of weightlifting dumbbells. The experiments were powered by a small nuclear generator.

Among the instruments was a **Lunar Passive Seismic Experiment** to detect moonquakes, whether natural or simulated by the crashing of space hardware, such as rocket stages or used lunar modules, on to the surface. The first impacts surprised NASA scientists because they set the Moon ringing like a bell for nearly an hour.

▼ Buzz Aldrin, photographed by Neil Armstrong looking over the Apollo 11 experiments.

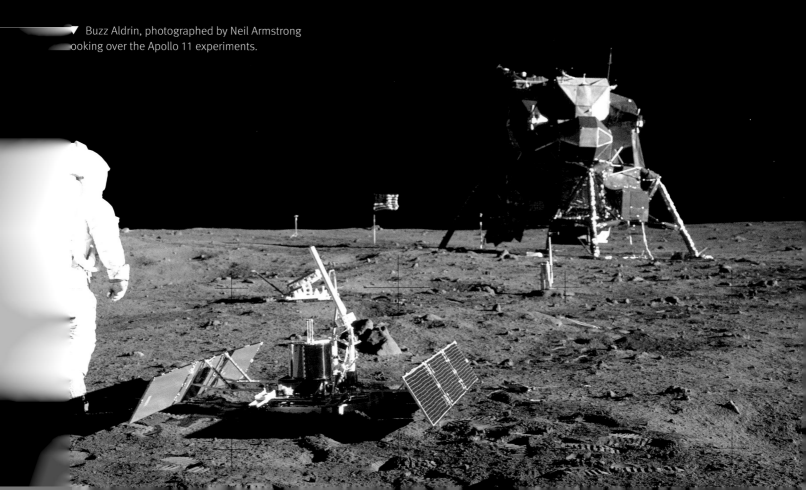

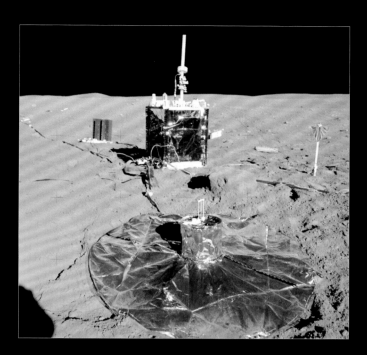

▲ ALSEP experiments on the Apollo 16 mission included the Passive Seismic Experiment in the foreground and an anchor flag for the Active Seismic Experiment at right. In the background are two power units.

A second device, the **Lunar Active Seismic Experiment**, was set to study the structure of the local site by measuring small explosions set up by the astronauts to create shock waves.

A **Heat Flow Experiment** involved drilling holes up to 2.3 m into the lunar regolith, into which instruments could be inserted to measure the temperature at several different points. It revealed that although the surface temperature varied by more than 250°C from day to night, subsurface temperatures remained constant, showing the lunar soil was an excellent insulator.

The **Lunar Surface Magnetometer** was built to measure local variations in the Moon's magnetic field. It has no global magnetosphere, unlike the Earth.

A **Lunar Dust Detector** measured the amount of moondust that accumulated on the surface. It was found to be very low.

One experiment in the package, the **Solar Wind Spectrometer**, was set up to measure the effect of the Sun's extended atmosphere. A separate experiment, a sheet of foil to trap particles in the solar wind, was brought back to Earth for analysis.

Not requiring power was the **Lunar Ranging Retro-Reflector**, a complex mirror aligned to face Earth and allow scientists to bounce laser beams, giving them a precise measurement of distance and demonstrating how the Moon's shape wobbles and distorts during its orbit.

Other experiments were set up to detect gravitational effects, meteoroid impacts, cosmic rays, charged particles and gases in the extremely tenuous lunar atmosphere.

SOVIET MOON MISSIONS

Having lost out in the race to put humans on the Moon, the Soviet Union attempted to take some of the gloss off the Apollo triumphs by sending unmanned spacecraft to the Moon. Several failed on the launch pad, including two missions aimed at returning lunar samples in the weeks before Apollo 11's historic landing. But on 20 September 1970, following the Apollo 11 and 12 landings and the drama of Apollo 13, a robotic probe called Luna 16 managed to land in Mare Fecunditatis (Sea of Fertility), drill into the rock and collect samples of it. The craft's ascent stage then fired to bring 101 g of lunar regolith back to Earth in a sealed container, landing by parachute in Kazakhstan.

The Soviets followed with another first, when Luna 17 landed the earliest remote-controlled rover on the Moon's surface on 17 November 1970. Called Lunokhod 1 (or 'Moonwalker 1'), the eight-wheeled vehicle trundled about inside Mare Imbrium (Sea of Rains) for 11 months, sending back thousands of images of the surface. A second rover, Lunokhod 2, reached the Moon aboard Luna 21 on 15 January 1973, landing in a crater called Le Monnier. It operated for four months, taking more than 80,000 images.

Two further missions, Luna 20 and Luna 24, successfully collected Moon samples also – Luna 20 from the Apollonius highlands near Mare Fecunditatis in February 1972, and Luna 24 from Mare Crisium (Sea of Crises) in August 1976. Other Soviet probes were put into orbit around the Moon.

▲ A model of the Soviet Luna 16, which became the first robotic probe to land on the Moon and bring samples of rock back to Earth, in September 1970.

THE APOLLO-SOYUZ PROJECT: A THAW IN RELATIONS

We have become used to collaboration in space between nations, in particular aboard the International Space Station. But the first example of such co-operation came back in 1975 between the Cold War superpower rivals: the USA and the Soviet Union. The result marked the final flight of an Apollo spacecraft, which linked up with a Soviet Soyuz in low-Earth orbit on 17 July.

The Apollo–Soyuz Project was being planned even as the final manned flights to the Moon were being made. And in April 1972, US President Nixon signed an agreement with Soviet leader Leonid Brezhnev for the two nations to meet in space, as a symbol of the thaw in their relationship.

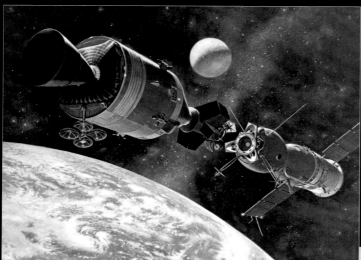

The Soyuz rocket launched first, from the Baikonur Cosmodrome, Kazakhstan, followed seven and a half hours later by the Apollo craft, atop the smaller of the Saturn rockets, the IB, from Florida. Commanding the Soyuz was veteran cosmonaut Alexei Leonov, with flight engineer Valeri Kubasov. The Apollo was commanded by Tom Stafford, on his fourth spaceflight, with first-timers Vance Brand and Deke Slayton. Slayton had been due to fly as one of the original Mercury astronauts but had been grounded by an irregular heartbeat.

The Apollo carried with it a special docking module that would allow the two non-compatible spacecraft mechanisms to link together. It was retrieved from the Saturn's third stage in the same way that the Lunar Modules had been before. As the Apollo approached the Soyuz, and fired small thrusters in its direction to brake, Leonov feigned alarm and joked: 'Tom, please don't forget about your engine.'

Three hours after docking, Stafford and Leonov shook hands through the Soyuz's open hatch. Then the two crews, who had spent much time together in the previous months, exchanged gifts, ate a meal together and carried out experiments. When the two spacecraft separated on 19 July to resume their individual missions, the Soyuz stayed in orbit another five days, and the Apollo nine.

There was drama during re-entry of the Apollo Command Module when poisonous propellant fumes were sucked into the capsule. As a result, the three astronauts spent two weeks in hospital in Honolulu after being picked up by the USS *New Orleans* on 24 July.

▼◄ The Apollo spacecraft is launched by a Saturn IB rocket from Florida (below) to rendezvous with the Soviet Soyuz that blasted off earlier from Baikonur to make the symbolic link-up in orbit around the Earth (as illustrated, left).

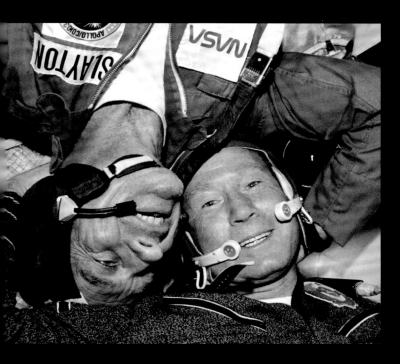

◀ Deke Slayton, left, and Alexei Leonov are all smiles as they greet each other in orbit, following the docking of their Apollo and Soyuz spacecraft.

SOYUZ – A RELIABLE WORKHORSE

The Russian Soyuz spacecraft was designed in the 1960s to succeed the Vostok that had carried Yuri Gagarin into orbit. It was meant to be capable of transporting cosmonauts on a mission around the Moon and back, but the Soviets lost interest in such a project after America got there first

with Apollo. Instead, the Soyuz has become the world's most successful crewed spacecraft, still delivering astronauts into space and back half a century after it first flew. In fact, following the retirement of America's Space Shuttle in 2011, it became the only way to ferry crews to and from the International Space Station for many years, as NASA and commercial companies continued working to produce their own spacecraft to replace the Shuttle.

The first crewed Soyuz launched on 23 April 1967, but the mission ended in tragedy when its parachute failed on re-entry, killing cosmonaut Vladimir Komarov. Another Soyuz mission, in June 1971, ended in disaster when the cabin depressurized before re-entry, killing crew members Georgy Dobrovolsky, Vladislav Volkov and Viktor Patsayev. Since then, the Soyuz has gained a reputation as a safe and reliable workhorse of the space programme.

The Soyuz has a basic modular form, with a front living area, or orbital module, attached to the re-entry capsule, and then the service module, containing engines, oxygen and fuel tanks, and other systems, at the rear. Though improvements have continually been made to the spacecraft, the basic design has stayed the same. A Progress cargo ship is also based on the design.

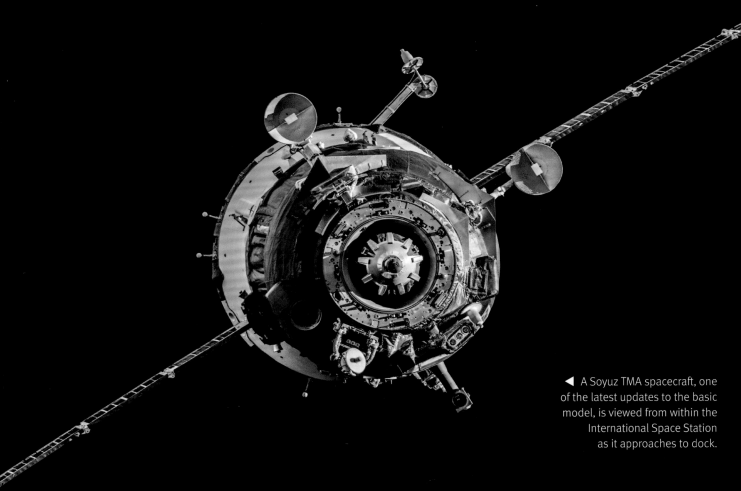

◀ A Soyuz TMA spacecraft, one of the latest updates to the basic model, is viewed from within the International Space Station as it approaches to dock.

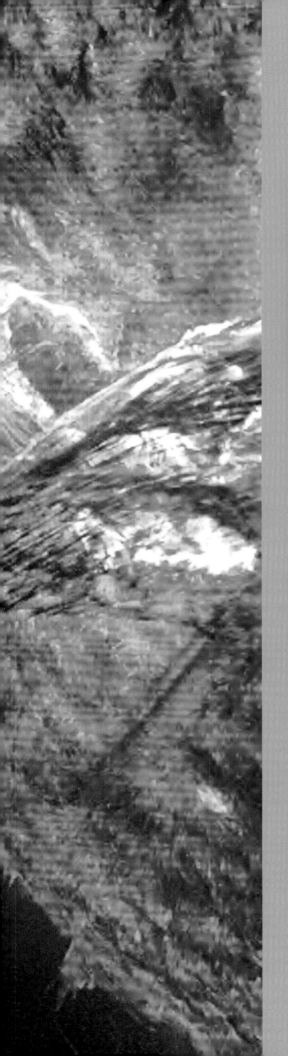

EXPLORING OUR NEIGHBOURS

A fiery prominence erupts from the Sun, a dramatic example of the violent activity that regularly occurs on this, our local star. Compared to many stars, the Sun is relatively benign and we would not exist without it. The Sun is orbited by a family of worlds, whose inner zone includes four rocky inner planets, a dwarf planet and a belt of debris left over from the formation of the Solar System.

THE SUN – OUR LOCAL STAR

The Sun is our local star, the powerhouse at the centre of the Solar System. It provides us with heat, light and a stable home, and is the only star that we can study close up. We know that the Sun is a ball of hot gas called plasma and 'burns' by a process of nuclear fusion, converting 600 million tons of hydrogen into helium every second. Despite being termed a yellow dwarf, the Sun is equivalent in volume to a million Earths. It is a typical type of star, about halfway through its life, with an age of around 4.5 billion years. If it had been burning like coal, it would have become an ember in a few thousand years! The Sun is on an evolutionary path that will see it swell in 5 billion years' time to become a red giant large enough to contain the entire 300 million km wide orbit of the Earth.

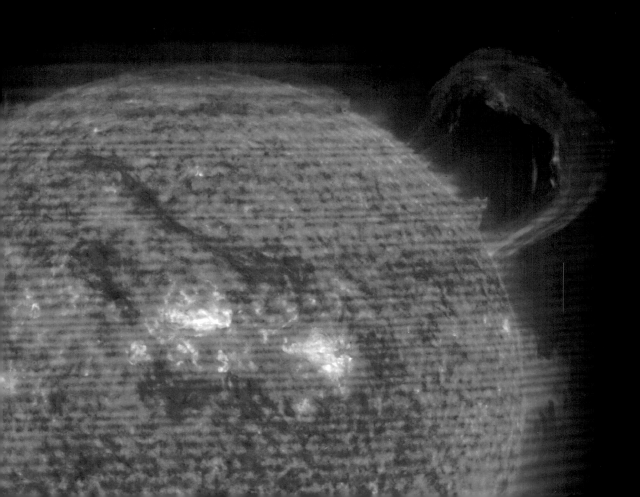

▼ A spectacular prominence – a huge cloud of relatively cool dense plasma, hanging in the Sun's hot, thin atmosphere, or corona.

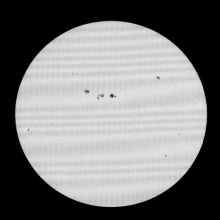

▲ A typical view of sunspots at a time of high activity on the Sun. This image was taken by the Solar Dynamics Observatory. The cluster at the centre is the width of about 14 Earths.

FEATURES ON THE SUN

Images of the Sun taken through a filtered telescope show a visible surface that is called the photosphere. This is not a solid surface but a layer about 100 km thick where the plasma suddenly becomes much less dense. It has a temperature of around 5,500°C.

Since early times it was noticed that occasional dark blotches would come and go on the photosphere. Some astronomers even suggested they were holes through which we might be able to see the Sun's inhabitants! We now know that these 'sunspots' are really cooler regions which only appear dark when compared to the rest of the Sun. Seen in isolation, they would glow brilliantly. Sunspots, which can become many times bigger than Earth, have a dark central region called an umbra and a lighter surrounding zone termed a penumbra. Like so many solar features, they are a product of powerful magnetic fields operating within the Sun, like iron filings drawn to a magnet. These fields are created by the movement of the plasma.

Other features visible on the photosphere include bright areas called faculae, where magnetic field lines become concentrated. And close-up images of the surface show a granular pattern of cells, each about 1,000 km across, where hot plasma bubbles up from within the Sun and then sinks again.

A SOLAR CYCLE

After astronomers began keeping regular, detailed records of sunspots in the mid-19th century, it was noted that their number varied over time. The Sun showed a cycle of around 11 years, with many more spots visible at the peak compared to the minimum. The last minimum, around 2008–9, was especially quiet with a completely spot-free Sun visible for most of the time. Historical observations indicate that the Sun underwent a prolonged quiet spell with very few spots between 1645 and 1715. During a solar cycle, the general location of sunspots shifts from the poles towards the

SUN – THE STATS

Sun's diameter at equator: 1.39 million km
Mass: 330,000 × Earth
Distance: 149.6 million km – the Earth's mean distance from the Sun of 149,597,870.7 km is termed 1 astronomical unit (au), and is used as a measuring stick to describe distances to more remote objects in the Solar System
Rotation period (day): 25 days at equator

equator as activity declines. Plotted on a graph, this migratory effect forms a pattern like a butterfly's wings. Spots tend to appear in pairs and the directions of their polarities switch with each 11-year cycle. Observations of spots showed that the Sun rotates once every 25 days near its equator, but in 35 days near the poles.

SOLAR ECLIPSES

A solar eclipse occurs when the Moon passes in front of the Sun. This can only happen at the phase New Moon, but does not occur at every New Moon because the Moon's orbit around the Earth is tilted against our own orbit around the Sun. By pure chance, the Moon appears the same size as the Sun in our sky, though it is really 400 times smaller but 400 times closer. The result is that when they do line up, the Sun can be completely covered for a few minutes for locations on a track across the Earth's surface, an event known as a total solar eclipse. For a much larger area around this track, a partial eclipse will be seen, with the Moon only partly covering the Sun's disk. During a total eclipse, the sky darkens, bright stars and planets appear, and the Sun's ghostly corona will come into view. At some eclipses, when the Moon is at its farthest from Earth, it does not quite cover the Sun's entire disk, leaving a ring of light around it. This is termed an annular eclipse.

▲ A view of the total solar eclipse of 21 July 2009, from Aitutaki Atoll in the Cook Islands, South Pacific Ocean.

INSIDE THE FURNACE

THE SUN'S LAYERS

We cannot see beneath the surface of the Sun, but seismological disturbances, or sunquakes, have allowed scientists to work out what lies within. With the help of computer modelling they have determined that the Sun's core makes up 20% of its diameter. Temperatures there reach 15 million degrees Celsius, with matter 150 times more dense than water. Around the core lies the radiative zone, extending out to 70% of the Sun's radius, and beyond that is the convective zone that lifts heated material to the surface, where it cools and falls again.

The photosphere is usually our visible surface, but telescopes fitted with hydrogen-alpha filters are able to see the Sun's inner atmosphere, or chromosphere, as a bright red band, about 10,000 km deep. Also in view, around the Sun's edge will be any flamelike prominences, spikes and other wisps of hot plasma sent soaring at high pressure from the photosphere. Prominences, which erupt and die over a period of many hours, may also be seen against the disk, when they are termed filaments.

The Sun's atmosphere does not end at the chromosphere, but extends millions of kilometres into space in a tenuous region called the corona, which reveals itself during a total solar eclipse and changes shape during the cycle of solar activity. One of the great puzzles in astronomy has been that the temperature within the corona soars to 2 million degrees Celsius, compared to 5,500°C at the surface. The exact mechanism for this is still being investigated, but it seems the heating may be caused by millions of tiny explosions a second on the Sun plus waves of energy flowing into its atmosphere.

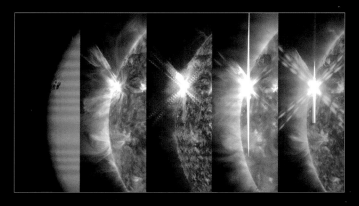

▲ NASA's Solar Dynamics Observatory monitored this flare on the Sun in several different wavelengths of light, including visible (seen at left), on 5 May 2015.

EXPLORING THE SUN

Countless space probes have been monitoring the Sun and space weather. Two early examples were Helios 1 and 2, a joint mission between NASA and Germany launched in 1974, which sent two probes to less than a third the distance of the Earth from the Sun to investigate solar plasma, the solar wind and its magnetic field.

Another angle on the Sun was achieved by Ulysses, a spacecraft launched from the Space Shuttle *Discovery* in 1990. Flying via Jupiter to take advantage of its gravitational pull, it settled into an orbit that carried it high above the plane of the Solar System, looking down on to the high solar latitudes, to learn more about the corona and heliosphere.

A major step in understanding the Sun came with the Solar and Heliospheric Observatory (SOHO), a joint mission

◄ An artist's impression of one of the twin STEREO spacecraft that have been giving scientists a 3D view of the Sun to help monitor space weather.

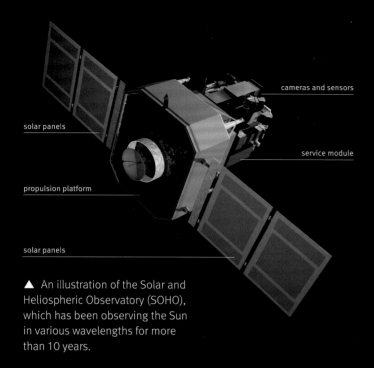

cameras and sensors

solar panels

service module

propulsion platform

solar panels

▲ An illustration of the Solar and Heliospheric Observatory (SOHO), which has been observing the Sun in various wavelengths for more than 10 years.

by NASA and the European Space Agency (ESA) to study our home star in a range of wavelengths. Launched in 1995, the spacecraft sits at about 1.5 million km from Earth, from where its cameras and instruments are informing scientists about the Sun from its heart to the outer corona and solar wind. As an unexpected bonus, SOHO has also discovered thousands of sungrazing comets!

Another major solar observatory is Hinode, a joint project between Japan, the US and the UK, which launched from Uchinoura Space Centre, Japan, in 2006. Hinode is Japanese for 'sunrise'. Three telescopes aboard watch for explosions on the Sun and their effects on its surroundings.

Also in 2006, NASA launched the Solar TErrestrial RElations Observatory (STEREO), a pair of spacecraft. One was placed well ahead of the Earth in its orbit and one left trailing far behind, to give a 3D view of the Sun and so help discover how coronal mass ejections occur and spread through space. After years of successfully monitoring space storms, contact was lost with the trailing probe, STEREO-B, in 2014, and engineers have since been trying to get it working again.

The latest major mission was NASA's Solar Dynamics Observatory, launched in 2010 into a geosynchronous orbit around the Earth, to learn more about the Sun's magnetic field, how it drives solar activity, and how that activity produces space weather.

Future missions include two launching in 2018 that will fly close to the Sun. ESA's Solar Orbiter will study the Sun's stormy surface and changes in the solar wind from an orbit that will take it to within 43 million km, or closer than inner planet Mercury. Even more daringly, NASA's Solar Probe Plus will fly right into the Sun's corona, to a point just 6 million km above the Sun's visible surface.

SUN FACT FILE

- Sunlight produced at the Sun's centre may take up to a million years to reach the surface, then less than nine seconds to reach the Earth.
- The invention of the spectroscope in 1814 by German optician Joseph von Fraunhofer allowed astronomers to discover what the Sun is made of, thanks to dark lines in the spectrum representing different elements.
- From a point on the Earth's surface, a total eclipse of the Sun can last from a few seconds to a maximum of 7 minutes 30 seconds.

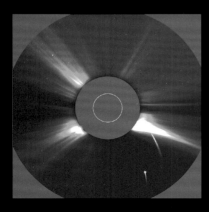

◀ Two comets appear in the field of view of one of SOHO's cameras. The glare from the Sun itself is masked out.

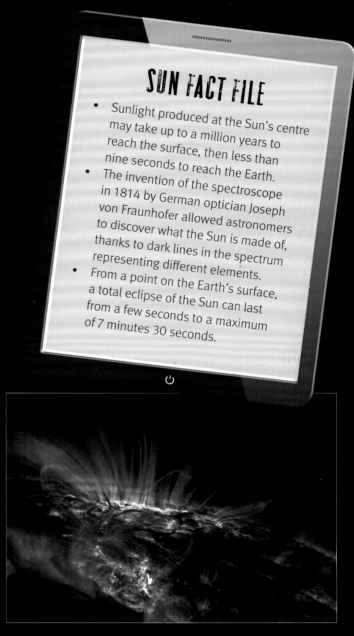

▲ Magnetically-driven structure is seen above a sunspot in an image by an early NASA solar observatory, the Transition Region and Coronal Explorer (TRACE).

SOLAR STORMS AND SPACE WEATHER

Apart from the sunshine that warms us and provides light, the Sun has a greater influence on the Earth and Solar System through the emission of charged particles, or plasma. This was demonstrated in a dramatic way in 1859 when amateur astronomers, including Richard Carrington in the UK, observed a brilliant flare on the Sun through their telescopes. A day later, the Earth was hit by what is now known as a geomagnetic storm. Brilliant displays of the aurora, or northern and southern lights, were seen from almost all of the Earth, and the telegraph communications system – an early kind of internet – was affected, with some stations catching fire or delivering electric shocks to their operators.

It became known as the Carrington Event. Flares like that observed by Carrington are produced by huge releases of energy from the Sun above sunspots, equivalent to billions of megatons of TNT. Monitoring of the Sun by early space observatories revealed that they are associated with even greater eruptions of energy that are now known as coronal mass ejections, or CMEs. These massive bursts can hurl vast quantities of electrically charged plasma and radiation out from the Sun – a phenomenon now termed space weather – and one or more must certainly have occurred to cause the Carrington Event.

If a CME occurs from the part of the Sun facing Earth, then the stage is set for phenomena such as the aurora. But as well as producing beautiful celestial light shows, major ejections aimed in our direction pose a threat to our modern infrastructure, including power grids, satellite electronics,

▼ A dramatic view of the aurora as seen from above, taken by ESA astronaut Alexander Gerst aboard the International Space Station in September 2014.

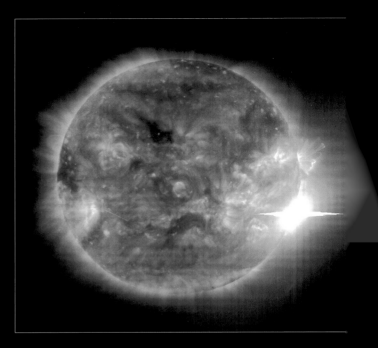

▲ One of the most powerful flares observed by the SOHO spacecraft was captured erupting in November 2003. The Sun is green as it represents an ultraviolet view.

and even the lives of astronauts in orbit. Flares and CMEs are far more frequent during the active phase of the Sun's cycle.

Apart from such dramatic, stormy events, the Sun is sending particles of ionized gas streaming far out into the Solar System all the time, at speeds of 400 km or more a second. It is a phenomenon known as the solar wind. Its existence was proposed during the 1950s, as evidenced by the shape and high temperature of the corona, plus the effect it had on the flimsy tails of comets. The earliest satellites and space probes confirmed the solar wind existed and that the Earth is effectively orbiting within the Sun's outer atmosphere. Large, cooler and less dense zones can occur in the Sun's atmosphere. Known as coronal holes, they allow a constant, rich stream of plasma to flow out into space.

▲ An artist's representation of how events on the Sun create space weather that impacts on the Earth and its protective magnetic field.

The effects of the solar wind and the Sun's magnetic field can be felt throughout the Solar System, far beyond the planets, and this region of influence is known as the heliosphere after Helios, the Greek sun god. The Earth has its own magnetic field, produced by the movement of liquid within the planet's core. It acts like a protective shield against the solar bombardment, steering the solar wind along its field lines so that dangerous radiation does not reach the Earth. The collision of charged particles from the Sun with atoms in the atmosphere produces the auroras seen around the Earth's magnetic poles. The more energetic the display, the farther from the poles it may be seen.

Right at the start of the Space Age, in 1958, two bands of charged particles, called the Van Allen Belts, were discovered by satellites to be held around the Earth by its magnetic field.

SAFETY WARNING

The Sun is so bright that it is extremely dangerous to look directly at it with any telescope or binoculars. Just a brief glimpse can damage your eyesight. Solar scientists and experienced amateur astronomers use professional-standard filters to study the Sun safely, or project the Sun's light through their telescope on to a sheet of card. **Beware** – unsuitable filters may appear to dim the Sun's light but still let through invisible, harmful radiation that can destroy your vision!

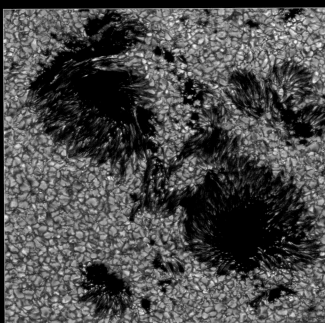

▲ A detailed view of a group of sunspots, taken by the Swedish 1 m Solar Telescope on La Palma in the Canary Islands.

◀ A powerful coronal mass ejection (CME) erupting from the Sun is observed by SOHO in December 2003. The Sun, masked out in the original image, is superimposed to show scale.

MERCURY – THE TWILIGHT ZONE

Mercury is the closest planet to the Sun and a rocky world that is slightly larger than the Moon. It was one of five planets, apart from Earth, known in ancient times because it may be seen with the unaided eye, but little was known about it until the first space probes were able to visit and found it was covered with craters. Mercury's fleeting appearances in the dawn and dusk skies led the Romans to name it after the winged messenger of the gods.

MERCURY REVEALED

Mercury is certainly fleet of foot, zipping around the Sun once every 88 days. The smallest, and innermost, of the four rocky terrestrial planets, with a diameter of 4,880 km, it is even smaller than Ganymede and Titan, the largest moons of planets Jupiter and Saturn respectively. Like our other inner neighbour Venus, it has no natural moon of its own.

Mercury is extremely dense for its size, a characteristic noted in 1841 by German astronomer Johann Franz Encke when he saw the pull it exerted on a passing comet. Radar observations have since showed that it has a very large and partly liquid iron core that makes up around three-quarters of the planet's diameter and more than 60% of its volume. Astronomers believe that either a planet-sized object struck Mercury billions of years ago, stripping away its outer layers, or they were vaporized in the extreme heat as Mercury was forming.

Mercury has a highly elliptical orbit that causes its distance from the Sun to vary from 47.01 million km when closest to 69.84 million km when at its most distant. At one time it was thought the planet might be always

► The first image of Mercury taken by NASA's Mariner 10 spacecraft in March 1974, showing a heavily cratered world. The black block towards the top is due to missing data.

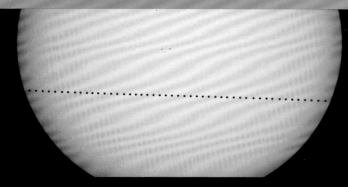

▲ A sequence of images superimposed shows how Mercury was observed to pass in front of, or transit, the Sun by the SOHO space telescope on 9 May 2016.

keeping the same face towards the Sun. However, we now know that it makes one rotation in a little under 59 Earth days. Three days pass locally for every two of its years. This slow rotation, coupled with the changing speed at which it travels in its orbit, means that the Sun appears briefly to reverse direction during sunrises and sunsets as viewed from parts of the planet's surface.

Life on Mercury is inconceivable since temperatures vary from being six times hotter than anywhere on Earth to more than twice as cold as our planet's coldest spot. There is no significant atmosphere. Instead Mercury is surrounded by a thin layer, known as an exosphere, of atoms of oxygen, sodium, hydrogen, helium and potassium that the solar wind and micrometeoroids have blasted from its surface. Water ice has been detected in permanently shadowed craters near the poles. Mercury has been found to have a tail, formed of sodium gas particles stripped away by the solar wind.

the twilight, but never look while the Sun is above the horizon as you could be blinded. Telescopes show that it displays phases like the Moon. The other way to spot it is when it passes in silhouette across the face of the Sun during a rare transit, but only attempt this if you have a telescope with a professional-standard solar filter.

MERCURY – THE STATS

Diameter at equator: 4,880 km
Mass: 0.055 × Earth
Rotation period (day): 58.6 Earth days
Mean distance from Sun: 57,910,000 km
Orbital period (year): 87.97 Earth days

TRANSITS OF MERCURY

Mercury's orbit around the Sun is angled slightly to our own, by 7°. That may not sound much, but it is enough to ensure that when the planet passes between us and the Sun it usually travels above or below it, rather than directly in front. Transits of Mercury across the Sun's disk are fairly rare and can only occur in May or November when the planes of our two planets' orbits intersect. These points are known as nodes. May transits occur at intervals of 13 and 33 years and November transits at intervals of 7, 13 and 33 years. The last transit of Mercury happened on 9 May 2016. The next will occur on 11 November 2019, 13 November 2032 and 7 November 2039.

OBSERVING MERCURY

You can see Mercury for yourself if you have a clear, unobstructed horizon to the east or west. It never strays far enough from the Sun to be seen in a dark, midnight sky, but when it appears farthest from the Sun, at a so-called elongation, it may be found either before sunrise or after sunset, resembling a bright star. The best time to look is when the ecliptic that marks the plane of the Earth's orbit is steeply inclined to the horizon. For mid-northern latitudes, this occurs on spring evenings and autumn mornings. Binoculars may help you to locate it in

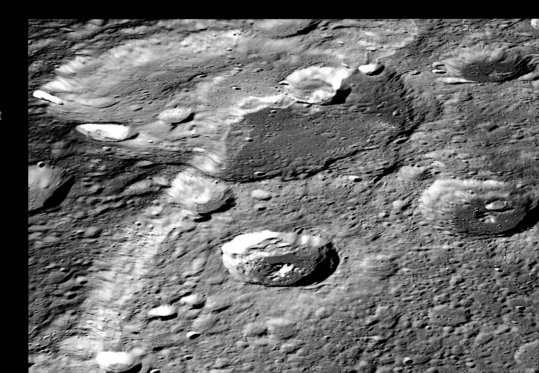

▶ A view from MESSENGER of the central region of Carnegie Rupes, a steep cliff produced by the cooling that caused Mercury to shrink. False colours indicate different elevations.

THE INCREDIBLE SHRINKING WORLD

Mercury has been visited by only two spacecraft: the NASA probes Mariner 10 and MESSENGER. They revealed that although Mercury is heavily cratered and thus superficially resembles the Moon, it has non-lunar features too, including numerous ridges lined by cliffs 1.5 km or more high. Its surface is darker than the Moon's, and the planet is thought once to have been encrusted with graphite. Mercury is also home to one of the biggest impact scars in the Solar System – known as the Caloris Basin, it is 1,545 km wide and was probably formed around 4 billion years ago. In the centre of the basin is a smaller crater, called Apollodorus, with more than 100 troughs radiating away from it like the threads of a spider's web. Surface features show that Mercury is shrinking.

▲ The colours in these scans of Mercury's surface by NASA's MESSENGER probe reveal the distribution of different minerals.

MARINER 10

Launched in November 1973, after humans had completed their visits to the Moon, Mariner 10 arrived at Mercury in 1974. The pull of the Sun makes it difficult to send a probe into orbit around one of the inner planets, so Mariner 10 was instead sent into an orbit around the Sun that would allow it to make occasional flybys of Mercury. It was the first spacecraft to use the gravity of another world – Venus – to reach its final orbit. Mariner 10 made three flybys of Mercury, in March and September 1974, and again in March 1975. More than 2,700 photos allowed it to take a close look at around 45% of the planet's heavily cratered surface. Mariner 10 confirmed there was no real atmosphere and discovered a magnetic field, showing Mercury had a large and iron-rich core.

MESSENGER

It was 30 years before NASA returned with MESSENGER – the MErcury Surface, Space ENvironment, GEochemistry and

▶ A natural-colour image of Mercury taken in 2008 by MESSENGER shows how it superficially resembles the Moon.

Ranging mission. Launched in August 2004, the space probe took a convoluted route, flying again past Earth and then twice past Venus to reach Mercury. MESSENGER then had to make three flybys of the planet, the first in January 2008, before it could finally go into orbit around it on 18 March 2011.

The probe carried a ceramic cloth sunshade and a cooling system to protect it from the intense heat. Though the mission at Mercury was planned to last a year, it was extended for a further three years, with a wealth of data and images of the planet being radioed back to Earth. But ultimately the propellant that MESSENGER needed to maintain its orbit ran out, and the probe crashed on to the planet's surface on 30 April 2015.

MESSENGER's seven years of observations boosted our knowledge of Mercury enormously. The probe mapped the entire planet, including the Caloris Basin in detail. Its images showed vents around the site that showed volcanism had shaped the planet as well as impacts. Looking down on the poles, MESSENGER confirmed there was water ice and found that some areas were covered by dark organic material. It also

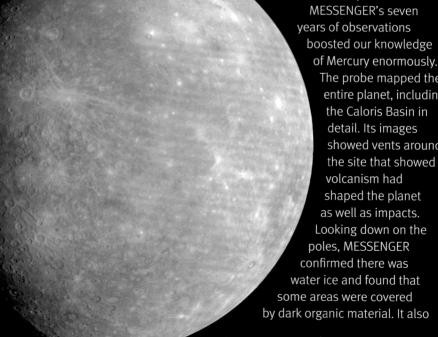

imaged thousands of depressions, dubbed hollows, which are something of a mystery. MESSENGER discovered Mercury's very tenuous exosphere and its tail produced by gas stripped away by the solar wind. Its images of huge wrinkly cliffs showed that the planet is contracting, which scientists attribute to the cooling of its giant core. The probe also found that Mercury's magnetic field is being produced by current activity within its core, and so is not a relic from the past.

BEPICOLOMBO

Mercury will next be put under scrutiny by a European-led mission called BepiColombo that is due to launch from Kourou, French Guiana, in 2017. The spacecraft will actually be formed of three components – two orbiters and a transfer module with an electric propulsion system. After a journey of seven and a half years, this module will be ditched and the orbiters will go into a polar orbit together. The information obtained when BepiColombo arrives will shed new light on the composition and history of Mercury, as well as helping to reveal how the inner planets in general formed, including Earth.

ESA is building one of the main spacecraft, the Mercury Planetary Orbiter (MPO), and the Institute of Space and Astronautical Science (ISAS) at the Japan Aerospace Exploration Agency (JAXA) will contribute the other, the Mercury Magnetospheric Orbiter (MMO). The MPO will

sun shield

Mercury Magnetospheric Orbite

Mercur Planetary Orbite

antenn

Mercur Transfer Modul

solar panel

▲ A diagram of the twin spacecraft, built by Europe and Japan, that make up the BepiColombo mission to Mercury.

study the surface and internal composition of the planet, and the MMO will study Mercury's magnetosphere, that is, the region of space around the planet that is influenced by its magnetic field.

The mission is named after Giuseppe (Bepi) Colombo, the 20th-century Italian mathematician whose work on spacecraft trajectories helped make previous Mercury probes possible.

▼ Mercury's natural colours are enhanced in this beautiful view from MESSENGER. Bright rays from meteoritic impacts are visible, plus dark rings around craters within the huge Caloris Basin feature, upper right. The backdrop image beneath is also from MESSENGER.

MERCURY FACT FILE

- Mercury is a rocky world that is a third the size of Earth, but bigger than the Moon.
- It makes two spins on its axis for every three orbits of the Sun.
- Craters on Mercury are named after deceased writers, musicians and artists.
- Temperatures on Mercury vary by 600°C, from 427°C on the Sun-facing side to −173°C on the night side.
- Mercury's axis is almost perpendicular to the plane of its orbit – as a result, it experiences no seasons.

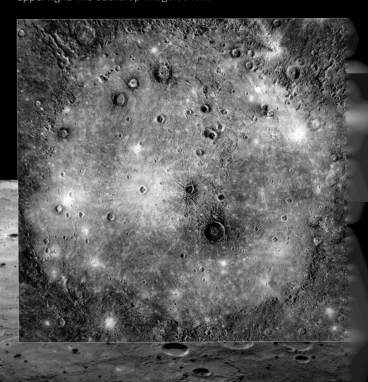

VENUS – A TWISTED SISTER

Venus, the second planet from the Sun, is another world known since antiquity. The brightest object in the night sky after the Sun and Moon, it stands out as a beacon that is impossible to miss. Its beauty led to our inner neighbour being named after the Roman goddess of love. But until the Space Age, nothing was known about what this rocky planet was really like because its surface is permanently shrouded in cloud. The truth is that Venus, of similar size to Earth, is a neighbour from hell.

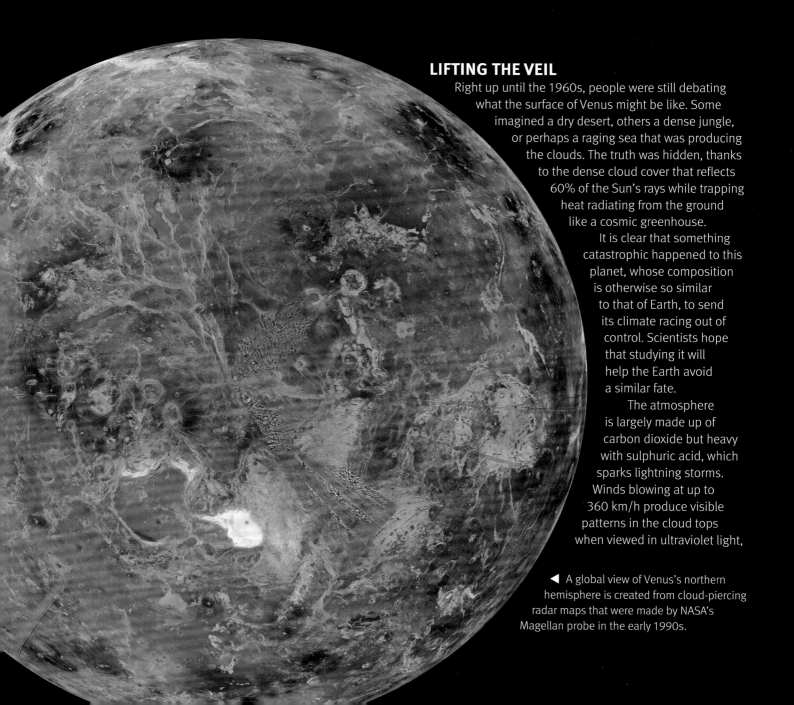

LIFTING THE VEIL

Right up until the 1960s, people were still debating what the surface of Venus might be like. Some imagined a dry desert, others a dense jungle, or perhaps a raging sea that was producing the clouds. The truth was hidden, thanks to the dense cloud cover that reflects 60% of the Sun's rays while trapping heat radiating from the ground like a cosmic greenhouse.

It is clear that something catastrophic happened to this planet, whose composition is otherwise so similar to that of Earth, to send its climate racing out of control. Scientists hope that studying it will help the Earth avoid a similar fate.

The atmosphere is largely made up of carbon dioxide but heavy with sulphuric acid, which sparks lightning storms. Winds blowing at up to 360 km/h produce visible patterns in the cloud tops when viewed in ultraviolet light,

◀ A global view of Venus's northern hemisphere is created from cloud-piercing radar maps that were made by NASA's Magellan probe in the early 1990s.

VENUS – THE STATS

Diameter at equator: 12,104 km
Mass: 0.815 × Earth
Rotation period (day): 243 Earth days
Mean distance from Sun: 108,200,000 km
Orbital period (year): 224.7 Earth days

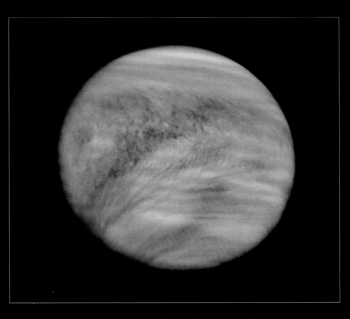

▲ Features in the cloud tops of Venus were revealed in this mosaic of images from the Mariner 10 probe in 1974.

and swirling anticyclones over both the planet's poles. Closer to the surface, however, the winds drop to a gentle breeze.

We know today that the surface of Venus is completely dry, and that it is about as inhospitable as one can imagine. The atmosphere is dense and poisonous with sulphuric acid, the air pressure is 93 times that found at sea level on Earth, and the temperature is around 470°C, or twice the maximum in a kitchen oven and hot enough to melt lead! Humans could not easily visit without being simultaneously poisoned, roasted and crushed to death.

Features on the surface were finally revealed thanks to cloud-piercing radar, both from Earth and from spacecraft, which allowed Venus to be mapped in some detail. Extensive plains are dotted with many thousands of volcanoes, believed still active, and their constant supply of magma means that Venus's landscape is regularly resurfaced over timescales of a few million years. Venus's continents have no seas to divide them now, but there may have been oceans billions of years ago, which simply boiled away into space.

JEWEL OF THE HEAVENS

Venus is so bright that it can actually be spotted in broad daylight in a clear blue sky. (As with Mercury, viewing should never be attempted with any optical aid if the Sun is also up as you could damage your eyes.) When Venus appears farthest from the Sun, at one of its elongations in the evening or morning sky, it can set up to five hours after the Sun or rise five hours before. At such times its light is dazzling and can cast shadows if you are at a site that is dark enough.

Using an early telescope at the start of the 17th century, the Italian astronomer Galileo noticed that Venus displayed phases just like the Moon, due to the fact it

▶ How the surface of Venus may look, with active volcanoes injecting sulphur dioxide into the atmosphere.

lies between us and the Sun. You can see these for yourself with just binoculars, but you will see no detail even with a telescope thanks to that cloud cover.

TRANSITS OF VENUS

Just like Mercury, there are occasions when Venus passes directly between the Earth and the Sun and is said to be in transit across the face of our home star. It then appears as a surprisingly large dark disk to observers with professional-grade filters to dim the sunlight. Transits of Venus in the 17th century from widely separated locations on Earth allowed the first measurements of the distances of Venus and the Sun to be determined. Sadly, unless you caught the transits of Venus on 8 June 2004 or 6 June 2012, you are unlikely ever to see one as they occur in pairs eight years apart but separated by more than a century! The next is not until 11 December 2117.

PRESSURE IS ON FOR PROBES

The Soviet Union made several attempts to study Venus from as early as 1961, but success did not come until 1970 when Venera 7 landed and sent back some weak signals. Venera 8 reached the surface in 1972, surviving for 63 minutes before being destroyed by the harsh conditions.

A number of subsequent Venera probes, built to withstand the heavy atmospheric pressure, had similar success over the next ten years, sending back photos of a rock-strewn landscape before their instruments melted. Then, in June 1985, two Vega probes on their way to Halley's Comet deposited balloons into the cooler Venusian clouds. They floated at an altitude of about 50 km, sending home data for 47 hours each.

NASA's first attempt to reach Venus was Mariner 1, which crashed into the sea after launch in July 1962. A month later, sister probe Mariner 2 launched successfully and flew past Venus in December 1962. Notable discoveries were Venus's backwards spin, its searing surface temperatures and air pressure, the mainly carbon dioxide constituency of the atmosphere, and its lack of a magnetic field.

NASA's next visit was with Mariner 10 which flew past Venus, picturing the wind patterns in its cloud tops, in February 1974 *en route* for Mercury. It followed up with another dedicated mission, Pioneer Venus, which saw two spacecraft arrive at the planet in December 1978. One was an orbiter, which carried out studies of the atmosphere and radar mapping of the surface, until it entered Venus's atmosphere and was destroyed in October 1992. The other was a multiprobe made up of a transporter and four separate

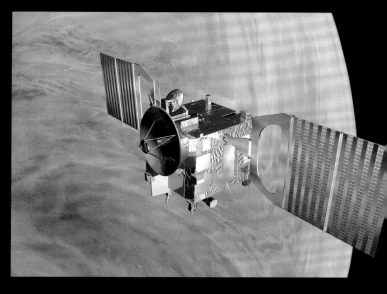

▲ An artist's impression of Europe's Venus Express orbiter during its highly successful mission.

probes fired into the atmosphere in November 1978, returning data for about one hour.

Extensive radar imaging of Venus followed in the early 1990s with NASA's Magellan mission, which went into a polar orbit around Venus in August 1990. It mapped 98% of the surface in high resolution, showing it was peppered with volcanoes, before contact was lost in October 1994. Two other NASA spacecraft — Galileo in 1990, and Cassini in 1998 and again in 1999 — observed Venus as they swung by, heading for Jupiter and Saturn respectively.

The European Space Agency (ESA) sent its first spacecraft to Venus in November 2005. Launched from Kazakhstan by a Russian Soyuz–Fregat rocket, this was a cut-price mission that reused parts designed for Mars Express, ESA's Mars probe. Venus Express studied the atmosphere in detail from a swooping orbit that brought it low over the cloud tops. It revealed huge variations in the amount of sulphur dioxide content, suggesting that the volcanoes were still active. Other clues included infrared images that closely resembled hot lava flows and unusually black terrain nearby, suggesting recent, unweathered volcanic deposits. Its fuel exhausted,

▼ Thermal imaging by Venus Express combined with radar mapping by Magellan provides strong evidence for active volcanoes.

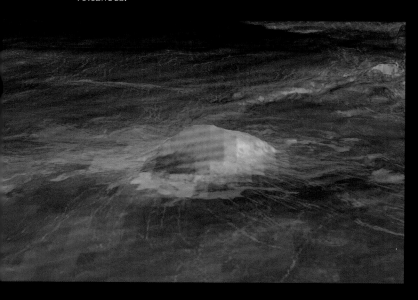

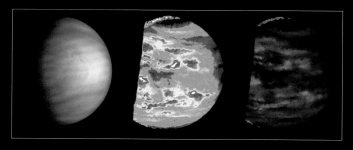

▲ Three colourized views of Venus by NASA's Galileo probe as it flew by to gain speed on its way to Jupiter, showing upper clouds, left, and infrared imaging at different levels of the night side.

◀ An image showing how a balloon might float for days in the clouds of Venus gathering data.

produce high-resolution maps of the surface topography and its composition. Another probe called DAVINCI (Deep Atmosphere Venus Investigation of Noble gases, Chemistry and Imaging) will descend slowly to the surface, studying the chemical make-up of the atmosphere.

But still more exciting missions are being proposed by planetary scientists for the long term to study Venus's atmosphere. One is for strong helium balloons to fly at varying heights gathering data. Even more ambitious is a semi-buoyant airship called VAMP (Venus Atmospheric Maneuverable Platform) that would float at the cloud base at night, then use solar power by day to soar back to the cloud tops. Some astrobiologists have speculated that simple microbial life forms might exist in the upper atmosphere, just above the cloud tops, where conditions are much more benign.

The Venus Express probe finally plunged into the hostile atmosphere in early 2015 after eight successful years.

Launched towards Venus in 2010, a Japanese space probe called Akatsuki looked lost after a fault caused it to fly past the planet. But five years later, mission controllers managed to rescue it and put it into a new, more elongated orbit where it began to survey the atmosphere.

FLIGHTS OF FANCY

NASA has shortlisted two Venus missions to fly in the early 2020s. The VERITAS (Venus Emissivity, Radio Science, InSAR, Topography and Spectroscopy) spacecraft is designed to

VENUS FACT FILE

- Venus has a day longer than its year. It orbits the Sun once every 224.7 Earth days, but a Venusian day is 243 Earth days long.
- It spins in a reverse direction to all the other planets apart from Uranus, which lies on its side.
- In the 1970s, the Arecibo Telescope in Puerto Rico revealed the presence of two continents on Venus, dubbed Ishta Terra and Aphrodite Terra, and a 11,000 m high mountain, named Maxwell Montes after 19th-century scientist James Clerk Maxwell.

How the VERITAS spacecraft will look in orbit around the cloud-shrouded planet Venus.

THE MOON – OUR COMPANION IN SPACE

The Moon is Earth's companion in space. Though smaller, its relative size makes our system almost a double planet, and it is taken for granted by most people as a serene, even romantic, addition to our skies. But the Moon had a dramatic origin and it is a very different world. Whereas the Earth is a blue marble, two-thirds covered by oceans, and teeming with life, the Moon is a dead, sterile and airless place.

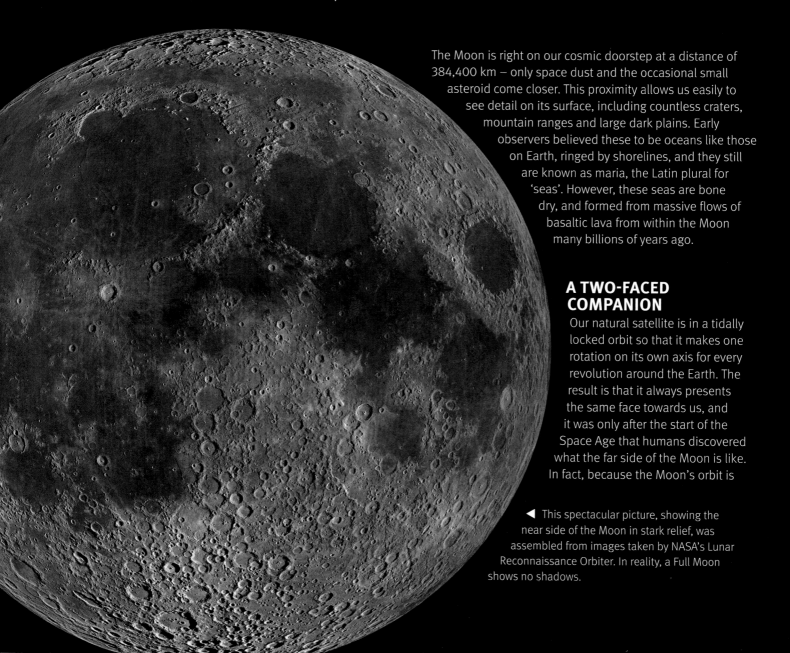

The Moon is right on our cosmic doorstep at a distance of 384,400 km – only space dust and the occasional small asteroid come closer. This proximity allows us easily to see detail on its surface, including countless craters, mountain ranges and large dark plains. Early observers believed these to be oceans like those on Earth, ringed by shorelines, and they still are known as maria, the Latin plural for 'seas'. However, these seas are bone dry, and formed from massive flows of basaltic lava from within the Moon many billions of years ago.

A TWO-FACED COMPANION

Our natural satellite is in a tidally locked orbit so that it makes one rotation on its own axis for every revolution around the Earth. The result is that it always presents the same face towards us, and it was only after the start of the Space Age that humans discovered what the far side of the Moon is like. In fact, because the Moon's orbit is

◄ This spectacular picture, showing the near side of the Moon in stark relief, was assembled from images taken by NASA's Lunar Reconnaissance Orbiter. In reality, a Full Moon shows no shadows.

THE MOON – THE STATS

Diameter at equator: 3,476 km
Mass: 0.012 × Earth
Rotation period: 27.3 Earth days
Mean distance from Earth: 384,400 km
Orbital period: 27.3 Earth days

THE MOON'S MAKE-UP

A glance through any telescope will show that the Moon is covered with craters. The lack of an appreciable atmosphere means that such ancient impact scars remain on view for eternity because there is no weathering as on Earth. The Moon (and Earth) got their biggest battering during a period in the Solar System's early history called the Late Heavy Bombardment. It would have been expected to attack the Moon evenly across its surface so geologists wondered why its far side is so different (see page 81).

▲ This image of the far side of the Moon, never seen directly from Earth, was captured by NASA's space-weather monitoring satellite DSCOVR in July 2015, as the Moon passed in front of the sunlit Earth.

not completely circular, it makes a slight rocking motion, called libration, that turns a little of the far side towards us along different parts of its circumference at different times. Over time, this wobble allows us to see about 59% of the lunar surface from Earth. Our first full views of the Moon's far side revealed that it is very different from the side that stays in view, being heavily cratered with few maria or large impact basins. The far side of the Moon is sometimes erroneously called the dark side, whereas the whole of the Moon experiences equal amounts of daylight and darkness during a single orbit of the Earth.

OBSERVING THE MOON

The lunar seas, or maria, can be made out with the unaided eye and help form the familiar face of the 'Man in the Moon'. Binoculars will reveal craters and a small telescope will allow you to study the Moon in considerable detail. The best time to look is when the line dividing the bright and dark sides of the Moon, known as the terminator, crosses the visible disk, since the shadows then make lunar features stand out in stark relief. Around Full Moon, the Sun's light is shining directly on to the Moon and the shadows disappear, but this is a good time to see the bright rays formed by material ejected in some ancient impacts. Occasionally at Full phase the Moon enters the Earth's shadow in space and a lunar eclipse occurs.

A VIOLENT BIRTH

Studies of lunar material brought back to Earth by Apollo astronauts and space probes have helped planetary scientists discover how the Moon formed. Latest research suggests there was a head-on impact between our world and an embryonic planet the size of Mars about 100 million years after the Earth formed, more than 4.5 billion years ago. The impactor, which has been dubbed Theia, threw out a cloud of debris that condensed to form the Moon. Comparing rocks from the Moon with those on Earth showed they have a similar composition, supporting their common origin.

➤ An artist's impression of the collision between an embryonic planet and Earth that is thought to have formed the Moon.

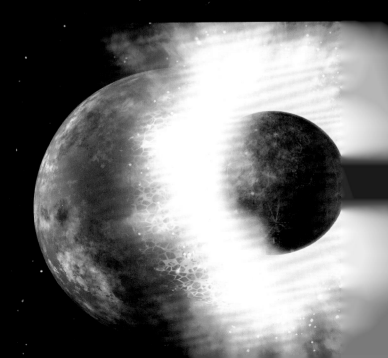

EXPLORING THE MOON

The human adventures of Apollo were a fading memory when the next phase of lunar exploration began, using unmanned spacecraft, a couple of decades later. Scientists wanted to know more about the Moon's interior, which Apollo's seismic and laser-ranging experiments showed is not uniformly solid, but has a syrupy centre.

Clementine was a joint NASA/Strategic Defense Initiative mission launched in January 1994 to orbit and map the Moon in detail, recording not just the shape of surface features but also their mineral composition. After 71 days in orbit, fulfilling its lunar mission by taking nearly 2 million images, the probe was intended to fly on to visit an Earth-passing asteroid called Geographos, but failed when a computer fault put it into a spin after it left the Moon's orbit.

Lunar Prospector was a low-cost NASA probe designed to learn more about what makes up the Moon's crust, measure its magnetic and gravitational pull, look for escaping gases, and locate water ice that was thought might lie within craters near the north and south poles which never receive any sunlight. Reaching the Moon in January 1998, the probe operated in a polar orbit for nearly 19 months before it was sent crashing near the south pole in a bid to detect the ice, though none was found at that time.

Europe's first mission to the Moon was SMART-1. Launched from the European Space Agency's spaceport at Kourou, French Guiana, in September 2003, its main purpose was to test the ion drive that propelled it, but it also carried out experiments to study the make-up of the Moon, how it formed and evolved, and the differences between the near and far sides. SMART-1 – it stood for Small Missions for Advanced Research in Technology – did not reach the Moon until November 2004, and its mission ended with a controlled crash into Lacus Excellentiae (Lake of Excellence) in September 2006.

A big advance in learning what lies inside the Moon came with NASA's GRAIL mission (Gravity Recovery And Interior Laboratory). This consisted of a pair of identical spacecraft, familiarly named Ebb and Flow, and each the size of a household washing machine, that flew in tandem around the Moon in polar orbits. Each carried an instrument that measured the distance between them with extraordinary precision, to just a few microns. This told them that the Moon's pull fluctuates across different regions, and the unevenness of its gravitational field helped explain why probes had previously been found to stray off course.

▶ A dramatic view of earthrise over the Moon, captured by NASA's Lunar Reconnaissance Orbiter spacecraft in October 2015.

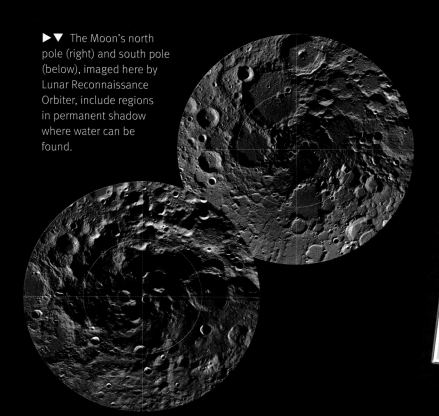

▶▼ The Moon's north pole (right) and south pole (below), imaged here by Lunar Reconnaissance Orbiter, include regions in permanent shadow where water can be found.

THE MOON FACT FILE

- The Moon is gradually moving away from the Earth at a rate of just under 4 cm a year.
- Both the Earth and Moon orbit around a point that lies about 1,700 km beneath our planet's surface.
- The Moon is the fifth largest moon in the Solar System and is bigger than dwarf planet Pluto.
- Gravity on the surface of the Moon is a fifth as strong as gravity on Earth.
- Temperatures on the Moon range from 107°C on the day side to −153°C when it is night.

GRAIL revealed that the lunar crust on the near side is thinner than on the far side. The side facing Earth has a mean thickness of 30 km, whereas on the far side it is up to 60 km deep. The thinner crust on the near side might explain why there are more 'seas', or maria, since the volcanic lava that filled them would have found it easier to reach the surface. GRAIL also located large, dense regions underground called mascons – short for 'mass concentrations' – which are thought to have been formed by giant asteroid impacts 4 billion years ago when the Solar System was like a vast shooting gallery. The twin GRAIL probes themselves collided with a lunar mountain near its north pole in December 2012.

NASA next turned its attention to the Moon's ultra-thin atmosphere, known as an exosphere. They wanted to see what formed a pre-sunrise glow that Apollo astronauts had reported seeing from space. The Lunar Atmosphere and Dust Environment Explorer (LADEE), launched in September 2013, flew low over the Moon to discover

that micrometeoroids hitting it are sending dust high above its surface, catching the light. LADEE was crashed into the far side in April 2014.

▶ The two GRAIL probes produced this gravity map of the Moon, where different levels of gravitational pull are represented by different colours.

▶ NASA's two GRAIL probes, Ebb and Flow, are pictured flying together in this artist's illustration.

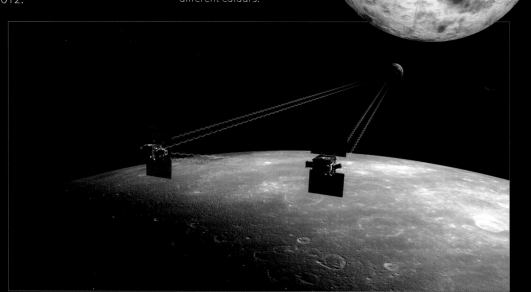

INTERNATIONAL LUNAR PROGRAMMES

WATER ON THE MOON

The Moon is clearly a dry desert of a world, and in the 1990s Lunar Prospector had failed to find any water ice. But it had been discovered, in a special form, locked away in basalt rocks brought back by Apollo 14 astronauts, and subsequent missions have revealed that there are indeed plentiful supplies. The difficulty will be in how to get at it!

India's first Moon mission, Chandrayaan-1 Lunar Orbiter, was launched in October 2008 carrying a NASA radar experiment to look for water ice. It found an estimated 600 million metric tons of it in the permanently shadowed regions of craters near the Moon's poles. Similar findings were made by two NASA spacecraft, the Lunar CRater Observation and Sensing Satellite (LCROSS) and Lunar Reconnaissance Orbiter (LRO), which had launched together in June 2009. LCROSS and its upper rocket stage were separately sent crashing into a crater called Cabeus in October that year and grains of water ice were detected in the cloud of debris ejected. LRO also found that nearly a quarter of Shackleton Crater, at the Moon's south pole, is covered with ice.

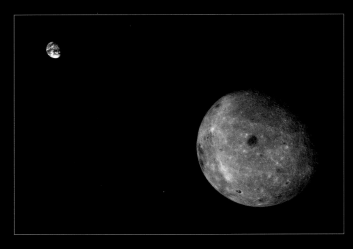

▲ This view of the Moon with the Earth in the distance was taken in October 2014 by China's Chang'e 5-T1 mission, a test flight in the country's lunar exploration programme.

If future missions are able to access the ice, it could be used by human colonizers not only to provide essential drinking supplies but also to help build space bases, and be converted into fuel to power rockets and other vehicles. Using local water will save on the expense of having to launch and ferry supplies from Earth.

In 2018, NASA plans to send a small CubeSat, piggybacking on a test flight of its new Space Launch System Moon rocket, to shine infrared lasers into the craters to find out more about their ice content. Two years later, a rover called Resource Prospector aims to carry out the first mining on the Moon by drilling up to a metre deep to extract samples.

CHINA'S MOON PROGRAMME

A major new player entered the field of lunar exploration in October 2007 when the China National Space Administration launched its first robotic probe, Chang'e, named after an ancient moon goddess, from the Xichang Space Centre in Sichuan province. After mapping and analyzing the lunar surface from orbit, it was crashed deliberately in March 2009. A follow-up mission, Chang'e 2, was launched in October 2010, but left its orbit in 2011 to begin heading for a new target, asteroid Toutatis, which it flew past successfully in December 2012.

In December 2013, China launched its most ambitious lunar mission yet, Chang'e 3, an unmanned craft that became the first since 1976 to make a soft landing on the Moon. As well as the Lunar Landing Vehicle, it was carrying a six-wheeled rover, Yutu, meaning 'jade rabbit', equipped with cameras and instruments to analyze the lunar soil, or regolith.

The landing site was a relatively fresh lava flow in Mare Imbrium (Sea of Showers), allowing scientists to study rock that was more like that found beneath the lunar crust. Yutu was still operating in early 2016, though no longer mobile.

▲ This strip-map of the Moon in three colours was made by an infrared instrument on India's Chandrayaan-1 probe, which surveyed the spread of different materials including water ice at high latitudes.

A further mission in the programme, Chang'e 4, is due to be launched some time before 2020 with another lander and rover, but this time to visit the far side of the Moon. It is due to be followed by a fifth mission, to collect at least 2 kg of rock and regolith from the Moon and bring it back to Earth. A test flight for this was performed when Chang'e 5-T1 was launched in October 2014 to fly around the Moon and then send a return capsule. This landed in Mongolia eight days later.

LAND OF THE RISING MOON

Japan's space agency, JAXA (Japan Aerospace eXploration Agency), launched its first Moon mission in January 1990 with a satellite, Hiten, named after a Buddhist musical angel. Hiten was put into a highly elliptical Earth orbit that sent it flying past the Moon ten times before it was deliberately crashed near the southern hemisphere crater Stevinus. Hiten released a separate probe, Hagoromo, but it is not clear whether it ever went into its own intended lunar orbit.

▶ A rocket launches Japan's first lunar orbiter, Kaguya, from the Tanegashima Space Centre in September 2007.

▼ China's Yutu rover is pictured exploring the lunar surface from a camera on the Chang'e 3 lander in December 2013.

Japan returned to the Moon in September 2007 with Kaguya, named after a fairytale princess, and earlier known as SELENE. It spend 20 months making a global survey to map the features, measure mineral deposits and make other readings before colliding, as planned, with a region of the near side that was in darkness. The flash was visible from Earth.

GOING PRIVATE

As well as the national space programmes, a number of private enterprises are aiming to reach the Moon. Google has sponsored a Lunar XPrize, with a $30 million purse, to encourage entrepreneurs to land a rover on the Moon, drive it 500 metres and send back video and images. And helped by some initial crowd-funding, Lunar Mission One is a UK-led bid by industry and educational institutions to send a robotic spacecraft to the Moon's south pole to drill and collect rock samples.

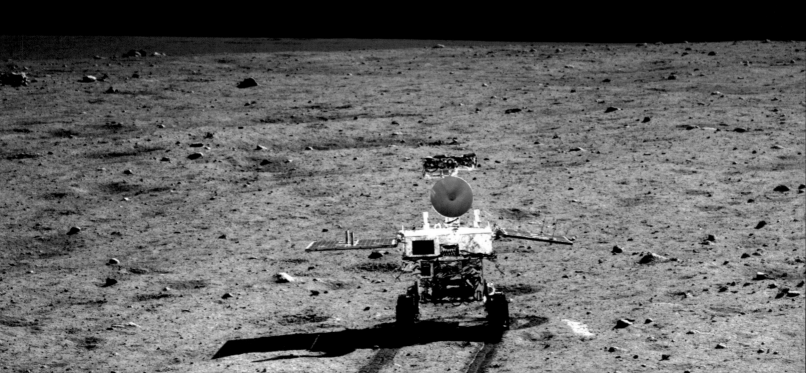

MARS – THE QUEST FOR LIFE

Mars, the fourth planet from the Sun, is another world known to the ancients. Its orbit brings it relatively close to Earth every two years or so, when it shines brilliantly in the night sky, then to the far side of the Sun when it appears considerably dimmer. The planet bears the name of the Roman god of war, and when at its brightest it does shine with a strong hue that might resemble a drop of blood – no wonder it also became known as the Red Planet. Though only around half the size of Earth, Mars is similarly rocky. Surface features observed with early telescopes led scientists and science fiction writers to wonder if it might be home to life.

▼ This striking image of Mars, a mosaic of 102 images taken by NASA's Viking 1 orbiter, is dominated by the spectacular Valles Marineris canyon, which is 3,000 km long and up to 8 km deep.

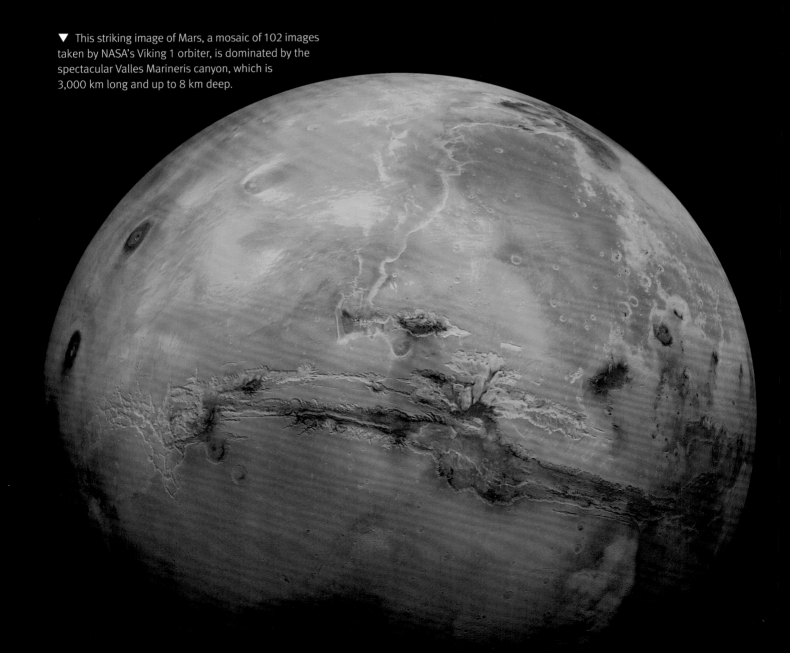

CHANGING VIEWS

Because Mars lies outside our own orbit, it can never show us crescent phases like Mercury and Venus. But at certain points in its orbit, it can take a gibbous shape, which Galileo noticed with his telescope. An early attempt to record surface detail was made by Dutch astronomer Christiaan Huygens in the 17th century. He drew a prominent marking that we today know as Syrtis Major. It allowed him to deduce, correctly, that a day on Mars, called a sol, is very similar in length to one on Earth. Huygens also made the first observations of a polar region on Mars by sketching a light zone around its south pole. In 1672, astronomers came to see that Mars had a tilted axis too and so experiences seasons just as Earth does. Other attempts to discern features on the Martian surface were less successful, but helped fuel the notion that Mars was inhabited. Observers believed they might be seeing seas, continents, clouds and patches of vegetation.

The idea that there were Martians got a major boost when Italian astronomer Giovanni Schiaparelli made a detailed map of Mars in 1877 and drew a network of fine lines that he called '*canali*', or channels. He won the support of leading US astronomer Percival Lowell, who had no doubt that these were water courses dug by aliens to irrigate their planet. The idea persisted well into the 20th century, but when spacecraft finally gave us close-ups of Mars, it was evident that the canals had been entirely illusory.

SEEING MARS FOR YOURSELF

Though Mars comes closer to us than any planet other than Venus, it remains in the neighbourhood for only a few weeks before becoming a remote object once again. But even that brief window offers amateur astronomers a chance to glimpse some of the more prominent features, such as the dark patch marking Syrtis Major, or the bright polar caps. Mars' disk shows largest when it is at opposition, the point when it lies on the opposite side of the sky to the Sun. This moment

▶ A photo of Mars taken with the Hubble Space Telescope shortly before the planet's close approach to the Earth in May 2016. Dark features surround the brighter Arabia Terra uplands at centre.

MARS – THE STATS

Diameter at equator: 6,794 km
Mass: 0.107 × Earth
Rotation period (day): 1.03 Earth days
Mean distance from Sun: 227,940,000 km
Orbital period (year): 686.98 Earth days
Number of moons: 2

always occurs within a few days of the planet being at its closest to Earth – the reason the dates do not coincide is due to the eccentric nature of Mars' orbit.

This eccentricity also means that Mars' distance from Earth varies enormously from one approximately two-yearly close encounter to another. In March 2012, for example, it came no closer than 101 million km, reaching a peak brightness (magnitude) of −1.2 and appearing just under 14 seconds of arc wide, which is about 120th the size the Moon appears to the naked eye. Since then closest approaches have become more favourable, and in July 2018 Mars will lie less than 58 million km from us, appear 24 seconds of arc across, and shine at magnitude −2.8. After that, Mars becomes more distantly separated again through the oppositions of October 2020, December 2022, January 2025 and February 2027.

Any telescope will show Mars' colourful hue, but you will probably need at least a 15 cm telescope to see any hint of its surface features. Occasionally, also, global dust storms blow up which hide detail from even the largest instruments. With your eyes alone, you can see the changing brightness of Mars at different times. When it is on a remote part of its orbit, it is hard to distinguish from a bright star, but at its closest it is a brilliant beacon. You will also notice that, for a few weeks around opposition, Mars appears to reverse direction in the sky. This phenomenon, called retrograde motion, happens when the faster-moving Earth 'overtakes' Mars and it appears to move backwards, just as a car does against more distant background objects when you pass it on the motorway.

◀ The HiRISE camera on NASA's Mars Reconnaissance Orbiter pictured these gulley channels in the southern highlands of Mars, evidence that water once flowed.

AN ANCIENT WATERWORLD

From the moment that the first passing spacecraft gave us close-up views of Mars, it was clear that the old ideas about our neighbouring world would have to be revised. Far from being a planet with vegetation or other similarities to Earth, it more resembled the Moon, with the first photos showing a rugged, cratered landscape. Subsequent missions have shown, however, that Mars is not like the Moon at all. While there is no sign of those imaginary canals drawn a century or so ago, surveys from orbit picked out many features that have clearly been carved out by water, such as former lake beds and long dried-up river gullies. Mars also has contrasting hemispheres. To the south are highlands, dotted with craters, while the north holds a vast low-lying plain that was probably produced by a collision with another large body 4 billion years ago. There is a huge canyon, like a gash in the crust, plus towering volcanoes that include the largest in the Solar System.

Visiting probes also confirmed that Mars has an atmosphere about 100 times thinner than Earth's. It also lacks a magnetic field, allowing ultraviolet radiation from the Sun to sterilize the ground, killing any likelihood of life there.

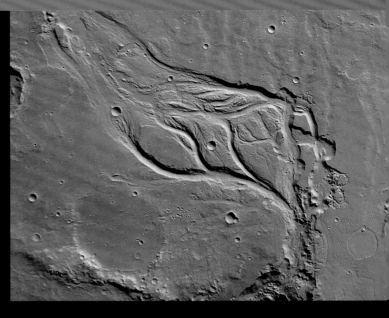

▲ Channels carved by the flow of water on Mars billions of years ago are picked out by Europe's Mars Express orbiter in a region called Osuga Valles.

SEEING THE SURFACE FOR REAL

We finally began to find out what Mars was really like with the first visiting spacecraft. In July 1965, NASA's Mariner 4 flew past. It grabbed just 21 photos. They were crude by modern standards, but they showed a landscape covered with craters. Mars looked dead, like the Moon, and there was definitely no sign of canals!

By chance, Mariner 4 passed over a particularly rugged region of Mars. Four years later, in 1969, the twin probes Mariner 6 and Mariner 7 flew by Mars, taking dozens of pictures. A chaotic and heavily cratered surface was again revealed as well as close-ups of the southern polar region. But the spacecrafts' flight paths meant that they too missed some of Mars' most dramatic features.

NASA's next successful probe, Mariner 9, transformed our knowledge of the Martian landscape because it went into orbit around Mars in 1971, becoming the first man-made spacecraft ever to circle another planet. As luck would have it, a global dust storm had blown up a couple of months earlier, obscuring the view, but the summits of Olympus Mons, the biggest volcano in the Solar System, and Mars' three Tharsis volcanoes were seen poking above the dusty cloud.

Apart from the volcanoes, the other big revelation from Mariner 9 showed Mars had a huge canyon stretching right across the planet, plus features that resembled dried-up river

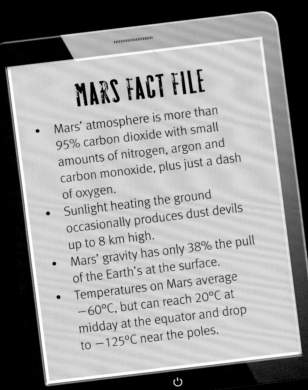

MARS FACT FILE

- Mars' atmosphere is more than 95% carbon dioxide with small amounts of nitrogen, argon and carbon monoxide, plus just a dash of oxygen.
- Sunlight heating the ground occasionally produces dust devils up to 8 km high.
- Mars' gravity has only 38% the pull of the Earth's at the surface.
- Temperatures on Mars average −60°C, but can reach 20°C at midday at the equator and drop to −125°C near the poles.

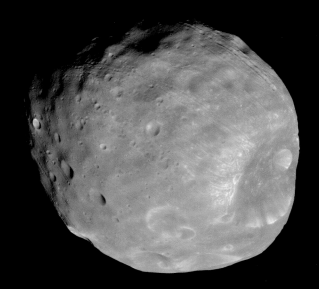

beds. There may not have been canals, but here were the first indications that water had once been flowing on the Red Planet. The vast canyon was named Valles Marineris after the space mission. It dwarfs the Grand Canyon in the US, being 4,800 km in length, up to 600 km wide and 8 km deep. Mariner 9 mapped the whole of Mars in more than 7,000 photos, and provided close-ups of its two tiny moons, Phobos and Deimos.

The Soviet Union also put two probes, Mars 2 and Mars 3, into orbit in November and December 1971. They sent back just 60 photos between them. Mars 2 carried a lander which crashed when its parachute failed. Mars 3's lander touched down, but communications failed after just a few seconds. Both Russia and NASA have suffered a number of failed bids to reach Mars.

FEAR AND PANIC

Mars has two moons, named Phobos (Fear) and Deimos (Panic). They are tiny compared to our own Moon, and are probably asteroids that were captured by Mars. They were first spotted in 1877 by Asaph Hall at the US Naval Observatory in Washington, though, coincidentally, Irish writer Jonathan Swift had predicted two satellites in his classic *Gulliver's Travels*. Phobos is larger, but still only 27 km across, and has a 9 km wide crater called Stickney. It circles Mars in 7 hours 39 minutes at a height of just 3,700 km above the ground. Latest research suggests that in several billion years' time, Phobos will be ripped apart by Mars' tidal forces, creating a ring of particles around the planet. Deimos is only 15 km across at its widest and takes 30 hours 18 minutes to orbit Mars.

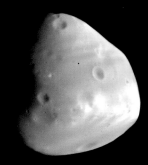

▲▶ Close-up views of Mars' two moons, Phobos (above) and Deimos (right), pictured by NASA's Mars Reconnaissance Orbiter. The depression formed by crater Stickney is visible on the right limb of Phobos.

METEORITES FROM MARS

Future missions to Mars will return samples of its rocks, but several have already been found here on Earth as a particularly rare form of meteorite. They were blasted out of the Martian crust by impacts billions of years ago and circled the Sun before colliding with Earth. Air trapped within them has been identified as matching the make-up of Mars' atmosphere in the ancient past. Controversially, some scientists have claimed to find fossils of Martian organisms too, notably in one picked up in the Antarctic, but most are highly sceptical.

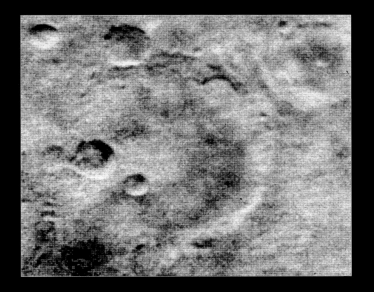

◀ One of the close-up photos from Mariner 4 in 1964 which revealed that Mars had craters.

▼ A panoramic view of the Martian surface from NASA's Pathfinder lander in 1997. It's little rover, Sojourner, can be seen inspecting a nearby rock in the centre of the image.

EYES IN THE SKY

The fascinating revelations from the Mariner probes spurred on interest in the Red Planet. A twin approach was made: a number of missions would be sent to orbit and survey Mars, while others would land and explore the surface. Orbital mapping has taught us that Mars was once a very different planet from the one we see today, with signs of vast ancient oceans and waterways nearly 4 billion years old. Mars also had a magnetic field and a denser atmosphere – but it lost much of its air, and the water disappeared into space or sank underground.

VIKINGS IN ORBIT

NASA's next missions after Mariner 9 were a pair of spacecraft, named Viking 1 and Viking 2, each combining an orbiter and a lander. Viking 1 arrived first, went into orbit in June 1976 and at once began to reconnoitre the surface below for a suitable site for the landers. This proved more tricky than expected due to the chaotic nature of the terrain, but a region called Chryse Planitia – 'the plain of gold' – was chosen for Viking 1's lander and Utopia Planitia for that of Viking 2. The orbiters mapped 97% of Mars in great detail, taking 52,000 images, and discovered that the north polar cap was primarily water ice rather than carbon dioxide, as had been believed. The Viking 2 orbiter operated until July 1978, while the Viking 1 orbiter lasted until August 1980.

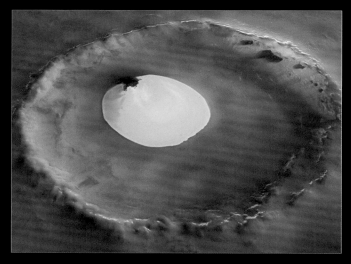

▲ A circular patch of water ice inside a 35 km wide impact crater on the plain Vastitas Borealis is imaged from the Mars Express orbiter, using its high-resolution stereo camera. More traces of ice can be seen on the crater's walls and rim.

MARS GLOBAL SURVEYOR

NASA began a new orbital study with Mars Global Surveyor. Launched in November 1996, the spacecraft took time to reach its circular near-polar orbit before beginning its primary mission to map the planet in March 1999. Its major findings included gullies and debris movement suggesting recent flows of liquid water, as well as evidence of a water-rich history. It observed regular weather patterns including dust storms and dust devils whipping across the surface, plus signs that Mars had a magnetic field billions of years ago. It also identified 20 impact craters that were created after its arrival. The probe operated until November 2006 when its battery died.

▲ How Mars may have looked 4 billion years ago with much of the planet covered by oceans.

MARS ODYSSEY

NASA's first mission of the new millennium was a resounding success. Arriving at Mars in October 2001, and named in honour of Arthur C. Clarke's most famous science fiction story, Odyssey is still orbiting the planet and sending back useful data. It carries a thermal imager to detect water and ice beneath the surface, and analyze and map the distribution of minerals, plus an experiment to study levels of harmful radiation that could affect future astronauts. Odyssey has also been operating as a communications relay for NASA's subsequent landers and rovers.

MARS RECONNAISSANCE ORBITER (MRO)

The next NASA probe sent to circle Mars was carrying the most powerful camera yet sent into the Solar System, which has returned highly detailed images of the Martian terrain. Arriving at Mars in March 2006, the probe began dipping into the atmosphere to slow down and enter its final orbit. It began scanning the surface to look for water features including the shorelines of ancient oceans and lakes. MRO is still working in 2016, and its detailed view has also picked out spacecraft on the ground, including the UK's Beagle 2 which failed to 'phone home' in 2003. It has also helped identify locations where future missions might land.

MARS EXPRESS

Europe's first mission to Mars was a low-cost spacecraft, built in a short time and launched by a Soyuz–Fregat rocket from Kazakhstan in June 2003. It reached Mars in December 2003, carrying seven instruments designed to study the Martian atmosphere and climate, and the geology and mineral composition of the surface and subsurface. Mars Express also had a high-resolution camera and sent back thousands of detailed views of the Martian terrain, many in 3D. Its radar also penetrated the surface enough to detect layers of water ice, including vast plains of permafrost and enough water in the polar caps to create a deep ocean across the whole planet. The spacecraft, which delivered the ill-fated Beagle 2, continues to operate in 2016.

MANGALYAAN

India launched its first interplanetary mission from Sriharikota in November 2013. Mangalyaan, which is Hindi for 'Mars craft', successfully went into Mars orbit in September 2014. The probe, also known as the Mars Orbiter Mission, carries five experiments to study the surface features of Mars, its mineralogy and atmosphere.

MAVEN

NASA returned to Mars orbit in September 2014 with the Mars Atmosphere and Volatile EvolutioN mission (MAVEN), which had launched the previous November. The probe's special task is to study the upper Martian atmosphere and includes a number of dives to get a better taste of the planet's air.

Scientists want to know more about how Mars' water and other gases were lost into space by the solar wind. First results show that the process is accelerated when storms rage on the Sun.

EXOMARS ORBITER

Europe headed back to Mars with the launch of a twin probe by the Russian space agency Roscosmos in March 2016. One module of the mission is the Trace Gas Orbiter, which will try to detect methane and other gases in the Martian atmosphere to help answer the question of whether there has ever been life on Mars. The other part of the mission is a test lander.

▲ An artist's impression of how NASA's MAVEN probe might look as it orbits Mars and studies its atmosphere. The spacecraft is designed to dive repeatedly to try to discover why the planet's once-dense atmosphere is being lost into space.

▶ Mars' south polar ice cap stands out in this sweeping and magnificent view across the heavily cratered highlands to the Hellas Basin (seen at top left) from Mars Express, imaged with its high-resolution stereo camera.

EXPLORING THE SURFACE

In the new millennium, NASA's focus has been on learning more about Mars' watery past, since that is key to knowing whether it might ever have been inhabited. Roving robots have begun studying areas once abundant with water to find clues as to whether there was ever life and, if so, what happened to it. They are also laying the groundwork for missions that will bring back rock samples and eventually human visits to set up bases on Mars.

▲ These tiny mineral balls dubbed 'blueberries' were imaged by the Opportunity rover in a patch of soil about 3 cm wide.

VIKINGS INVASION

Having separated from their orbiters, the two Viking landers became America's first spacecraft to touch down safely on Mars. From their stationary locations, they not only took more than 1,400 photos of the Martian surface, revealing a rock-strewn, sandy landscape, but they also acted as simple laboratories. In a bid to find signs of life, samples of the soil were grabbed, heated and analyzed. There was a violent and unexpected reaction when nutrients were added, which left planetary scientists puzzled. The majority view, however, was that the reaction was chemical rather than biological, with Mars' topsoil too sterile for present-day life, due to the bombardment of ultraviolet radiation from the Sun. Viking 2's lander stopped communicating in April 1980 and Viking 1's lander ceased in November 1982.

PATHFINDER TO MARS

After Viking, it was 20 years before the next successful excursion to the Red Planet. NASA employed a new landing technique when they used airbags to bounce their first robotic explorer on to the Martian surface on Independence Day, 4 July 1997. The Pathfinder lander released a rover, Sojourner, to run about its landing site on an ancient flood plain called Ares Vallis. In three months of operation, the mission captured more than 17,000 images, sampled rocks and soil and weather conditions, and found that Mars was once much warmer, with a thicker atmosphere and extensive surface water.

SPIRIT AND OPPORTUNITY

In January 2004, two of NASA's most famous Mars explorers bounced to landings on opposite sides of the planet, using the airbag technique that worked for Pathfinder. Spirit and Opportunity were Mars Exploration Rovers, designed to operate initially for 90 days each. Both greatly exceeded their engineers' best expectations. Spirit, in the 150 km wide former lake Gusev Crater, kept on trucking until it became stuck in soil in May 2009, and contact was lost in March 2010. Opportunity was incredibly still operating around its landing site in Meridiana Planum in 2016, more than 12 years after arriving on Mars.

Both rovers found mineral deposits that showed Mars was once awash with water. Opportunity imaged haematite spheres, dubbed 'blueberries' by scientists, which are mineral balls left behind by water, like a ring around a bathtub.

PHOENIX

The US's next visitor to the surface of Mars, in May 2008, was a simple lander which NASA sent to the far north to check for water and monitor polar weather. Phoenix used parachutes and thrusters to soft land on the vast plains of Vastitas Planum above the 65th parallel. Phoenix – so named because it was resurrected

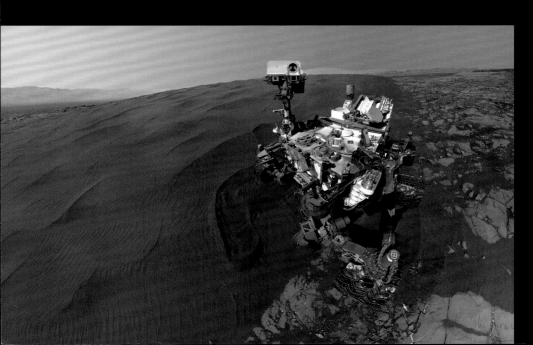

◄ A self-portrait of Curiosity in 2015, assembled from several different photos. The arm holding the camera moved between shots and so does not appear in the final mosaic.

from a cancelled mission – discovered a terrain like our own Arctic tundra, with polygonal patterns due to expanding and contracting ice beneath the surface. On landing, the probe found itself to be sitting on a slab of ice uncovered by its thrusters during descent. A scoop of soil also seemed to contain ice, which later evaporated into the air, or more correctly sublimated as there is no liquid stage between ice turning to vapour on Mars. Phoenix detected the first snowfall observed on another world, and also discovered a chemical compound called perchlorate, which on Earth is toxic to life. As winter set in, Phoenix succumbed to the falling temperatures at its inhospitable site and died in November 2008, becoming entombed in ice.

MARS SCIENCE LABORATORY (CURIOSITY)

NASA returned to the rover principle for its next mission, known as Curiosity for short. The vehicle was the size of a small car and three times heavier than a Mars Exploration Rover like Opportunity, so airbags could not practically land it. Instead a daring technique called the Sky Crane manoeuvre was used. After parachuting to subsonic speeds, a descent stage detached itself and fired eight thrusters to hold itself above the planet's surface, before lowering Curiosity gently by tether to the ground inside Gale Crater. The landing was dubbed 'seven minutes of terror' and, incredibly, it worked! Curiosity touched down on Mars in August 2012 and has been exploring ever since around the central peak, Mount Sharp. The vehicle is a complete mobile laboratory that has made several discoveries. It identified an ancient riverbed, measured levels of radiation, found traces of methane, and determined that the landing site could have supported microbial life billions of years ago, though none has actually been found.

▼ This formation of rock strata pictured by Curiosity inside Gale Crater indicates that water once flowed through the area.

EXOMARS

As its companion orbiter studies Mars from above, a demonstration lander, named Schiaparelli after an early observer of the planet, is due to carry out a test run for the European Space Agency in October 2016. If all goes well, a second ExoMars mission will launch in in 2020 to land a rover that can drill into the planet and search for signs of ancient life.

INSIGHT – POSTPONED

America's next lander, InSight, will be a stationary platform designed to drill up to 5 m deep into the Martian crust. It will measure heat from within to tell whether the planet's core is solid or liquid, while another instrument will detect marsquakes. It was due to launch in March 2016, but the mission has been delayed at least two years after a leak was detected in one of its main experiments. The probe's name stands for Interior Exploration Using Seismic Investigations, Geodesy and Heat Transport.

▼ An ExoMars rover that the European Space Agency is sending to explore more of Mars.

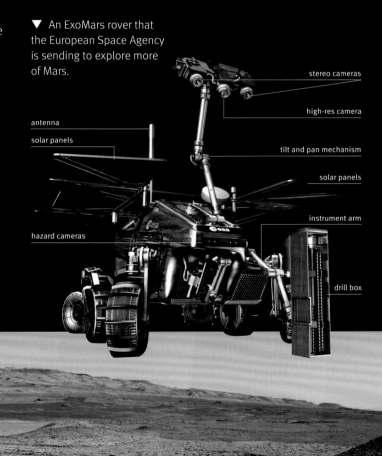

stereo cameras

high-res camera

antenna

solar panels

tilt and pan mechanism

solar panels

instrument arm

hazard cameras

drill box

CERES AND THE ASTEROID BELT

Scattered across the Solar System, but mainly concentrated in a belt a third of the way between Mars and Jupiter, are the asteroids. They are leftover debris from the formation of the Solar System, and might well have collected to make another planet had Jupiter's powerful gravitational pull not prevented that happening. Largest object in the zone is Ceres.

Space probes have paid visits to a number of asteroids. The most ambitious has been NASA's Dawn mission, which spent more than a year surveying the large asteroid Vesta before heading off to orbit Ceres, which has been classified as a dwarf planet since 2006.

HUNT FOR THE MISSING PLANET

Before any asteroids were known, astronomers in the 18th century had become convinced that the gap between Mars and Jupiter was too large and that there must be a yet-undiscovered planet lurking there. Hungarian nobleman Baron Franz von Zach organized a hunt

◀ A view of dwarf planet Ceres in natural colour as taken by NASA's Dawn spacecraft in May 2015. Bright spots are visible in two craters.

for it in 1800, allocating different observers a region of sky to search. They became known as the Celestial Police.

On New Year's Eve 1800, a hitherto unrecorded object was identified in the constellation of Taurus, though not by one of the Baron's team of searchers. A Sicilian monk, Giuseppe Piazzi, spotted it from Palermo Observatory and followed its movement for several weeks. He had discovered Ceres.

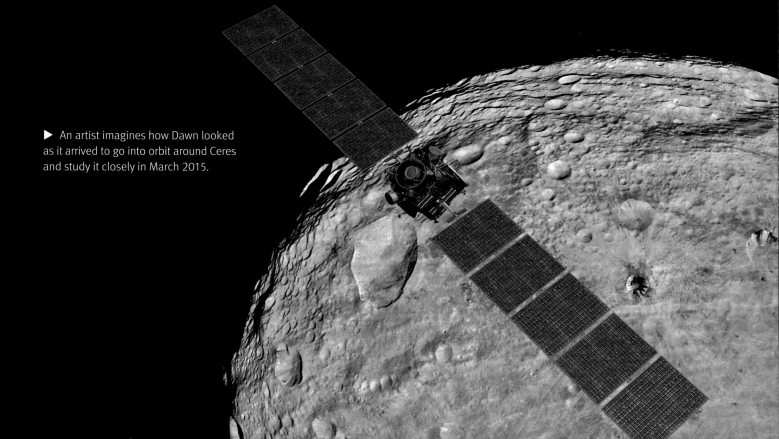

▶ An artist imagines how Dawn looked as it arrived to go into orbit around Ceres and study it closely in March 2015.

If this was the missing planet, however, it was not alone. In 1802, Heinrich Olbers, of the Celestial Police, located a second object, which was named Pallas. Karl Harding found a third, Juno, in 1804, and Olbers discovered his second, Vesta, in 1807. They were called planets at the time, but, since it was clear they must be a lot smaller than normal planets, leading astronomer of the time William Herschel came up with the term asteroid, meaning starlike, to describe them.

Hundreds of thousands of asteroids have been discovered since then, with 200 or so bigger than 100 km across and 750,000 larger than 1 km across. Though the vast majority are Main Belt asteroids between Mars and Jupiter, some are grouped together and locked in locations relative to the orbits of various planets. Others have been steered into orbits that bring them across the orbit of Earth and they are known as Near Earth Asteroids, or Potentially Hazardous Objects because of the slight risk that one could collide with our planet.

Asteroids come in three main types – carbonaceous (rocky clay), silicaceous (a stony-iron mix) and metallic. A number have been found to have smaller asteroids orbiting around them, and one even has its own set of rings. But if you added all the asteroids together, you would not have enough material to make an object the size of our Moon.

CERES – THE STATS

Diameter at equator: 945 km
Mass: 0.00015 × Earth
Rotation period (day): 9 hours 4 minutes
Mean distance from Sun: 413,690,250 km
Orbital period (year): 4.6 years

HARVESTING DATA ON CERES

The largest asteroid, Ceres, named after the Roman goddess of harvest, won its new dwarf planet status because it had achieved a spherical shape. Hubble images showed that, and suggested there were mysterious bright regions on its surface. Little more was known until NASA's Dawn mission.

Arriving from Vesta, Dawn went into orbit around Ceres on 6 March 2015 to begin a detailed survey. *En route*, as the little world began to loom larger, scientists became intrigued by bright spots that stood out against the general tone of the surface. From orbit, Dawn's photos showed that the brightest were within a crater called Occator and that there was a cluster of them. But were they ice or salt? The general view is that they are the highly reflective salt residue left after briny water burst from within Ceres and the water escaped as vapour into space.

Ceres could contain more water than in all the Earth's oceans, according to findings by ESA's Herschel Space Observatory, but in the form of ice beneath a thin crust and wrapped around its rocky core. This dwarf planet contains a quarter of the entire mass in the asteroid belt.

Another peculiar feature on Ceres is a conical mountain, about 6 km high, and with bright streaks on its flanks. It has been named Ahuna Mons. Scientists are having trouble explaining how it came to be there.

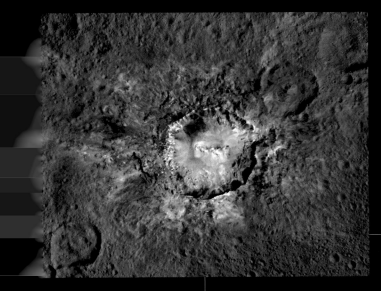

Bright features are seen close up in Ceres' 34 km wide Haulani Crater, as well as signs of landslides from its crater rim.

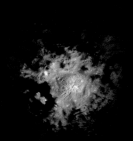

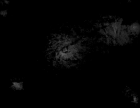

The cluster of bright spots in the crater Occator on Ceres are imaged in detail from the Dawn probe. Scientists believe they are salt.

VESTA AND THE REST

Before flying on to Ceres, the Dawn spacecraft made an extensive study of Vesta, spending 14 months in orbit from July 2011 to September 2012. Vesta is more typical of an asteroid, being a mainly rocky body – quite different from Ceres with its icy mix.

Vesta is the largest of the ordinary asteroids, and also the brightest in the sky. In theory, it should just be visible with the naked eye under exquisitely clear dark skies. Intriguingly, scientists already have some samples of Vesta to study here on Earth, because analysis of a type of meteorite – a stone from space that survives the fall intact – has identified it as coming from the asteroid, after material was blasted out of it by an ancient impact.

Before Dawn arrived, we only had views from the Hubble Space Telescope showing its rough shape. Dawn changed all that. Its wealth of images showed that the asteroid has an ancient surface covered with impact craters.

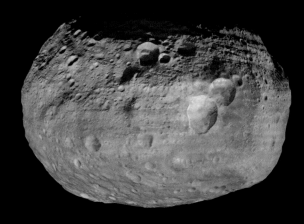

▲ Images of Vesta taken by Dawn have been combined and had colours added to indicate the variety of minerals found on the asteroid's surface.

MISSIONS TO THE ASTEROIDS

The first spacecraft to visit the asteroids was NASA's Galileo on its way to study Jupiter. It flew by Gaspra in October 1991, giving astronomers their first close view of one of these minor bodies, and then asteroid Ida, in August 1993, revealing that it was being orbited by a tiny moon, Dactyl.

The same agency launched a probe called NEAR Shoemaker (Near Earth Asteroid Rendezvous) in February 1996 to study Eros, an asteroid that regularly crosses our own orbit. It became the first spacecraft to orbit an asteroid, in February 2000, and land on its surface a year later. *En route* it also flew by another asteroid, Mathilde, in June 1997.

A number of other asteroids have been visited by spacecraft on their way to other missions. Europe's Rosetta checked out Šteins in 2008 and Lutetia in 2010 as it headed for its main target, a comet. Others include China's lunar mission Chang'e 2, which flew to within 3.2 km of Toutatis in 2012. Scientists have also used radio telescopes on Earth to map close-passing asteroids by bouncing radar signals off them.

► An artist's concept of NASA's OSIRIS-REx spacecraft arriving at asteroid Bennu to collect a sample before flying it back to Earth.

FALCONS SEEK THEIR PREY

Samples of an asteroid were brought back to Earth by a Japanese spacecraft called Hayabusa. Launched in May 2003, it reached an asteroid called Itokawa in 2005 and touched down twice on the surface, collecting a small amount of dust. The sample was delivered back to Earth by a capsule in June 2010. A small companion probe called Minerva was intended to hop around on the asteroid but failed.

Japan launched a second, more ambitious, sample-return mission in December 2014 to another asteroid, called Ryugu. The ion-engined craft is due to reach its target in June 2018 and survey it for 18 months, during which time it will plant a European lander called Mascot and a rover, Minerva 2, on the surface. Finally, a little missile will be fired to create a small crater. Then the spacecraft will try to touch down on the asteroid itself to collect material exposed by the impact and send the samples back to Earth.

UNITED STATES SEEKS OWN SAMPLE

Due to launch in late 2016 is NASA's own mission to grab a bit of an asteroid. Its name is a bit of a mouthful – the Origins-Spectral Interpretation-Resource Identification-Security-Regolith Explorer – but it is known as OSIRIS-REx for short. Two years after launch, the probe will rendezvous with a 250 m wide asteroid, Bennu, which has an orbit that brings it close to Earth. After mapping it for a year, and also studying forces that affect its potentially threatening orbit, the probe's arm will grab a small amount of ancient material from its surface that will be landed by capsule back on Earth in 2023. Bennu is thought to be rich in carbon, and may help scientists learn about how the ingredients for life were formed as the Solar System was born. OSIRIS-REx will also help prepare for future crewed missions to asteroids. Some commercial companies are planning ventures to mine these rocks of their precious resources.

▶ Gaspra, the first asteroid to be visited by a space probe, as seen by Galileo in 1991.

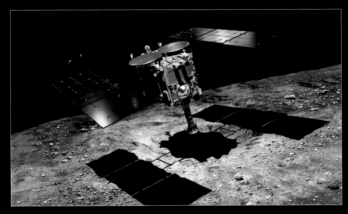

▲ Japan's Hayabusa 2 moves in to remove material from asteroid Ryugu which it will also fly home for scientists to study.

▼ A graphical comparison of the sizes of Ceres (left), Vesta (centre) and other assorted asteroids (named at right) that have been visited by spacecraft in recent years.

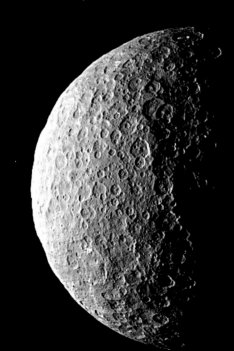

Lutetia

Mathilde

Ida

Eros

Gaspra

Šteins

MAKING
SPACE HOME

NASA astronaut Mike Hopkins floats against the blue backdrop of planet Earth as he makes a spacewalk to carry out repair work on the International Space Station, an orbiting laboratory that has been home to visitors from many nations. This colossal structure, which has allowed humans to maintain a permanent presence in space, is the culmination of years of effort by space agencies to master low-Earth orbit.

LIVING IN SPACE

The start of a new era of humans living in orbit came with the building of the first space stations, forerunners of the International Space Station that now circles the Earth. Though much smaller, they incorporated docking mechanisms, airlocks and room for their astronaut occupants to live and work.

THE SOVIET SALUTE

Once again, the Soviet Union led the way by launching a series of orbital outposts under the Salyut name, which means 'salute'. Though they shared the name, two different designs were produced, one for the civilian programme, and another for secret military activities, which was otherwise known as Almaz, or 'diamond'.

Salyut 1 was launched on 19 April 1971, followed by its crew in Soyuz 10, but they were unable to dock properly with the station and returned to Earth. Soyuz 11 launched on 6 June and successfully delivered the station's first crew – Georgy Dobrovolsky, Vladislav Volkov and Viktor Patsayev. Their 23-day stay went to plan, despite a small fire on day 11. However, the mission ended tragically when the three men died of suffocation after a pressure valve opened, allowing air to escape their capsule as it was about to re-enter Earth's atmosphere.

Salyut 1 was brought down to burn up over the Pacific in October after 175 days in space. Six more successfully launched, but there were also two failures. They increasingly gave cosmonauts experience of long spells in space. The last of these space stations, Salyut 7, launched on 19 April 1982

▼ Smiling cosmonaut Leonid Kizim makes a spacewalk during a mission in 1986 to visit both the Salyut 7 and Mir space stations.

and spent 8 years 10 months in orbit. The station shut
itself down while unoccupied in February 1985, but a
daring mission, Soyuz T-13, was sent to repair it and return
it to use. The final mission to Salyut 7, by another Soyuz,
was made to collect equipment and transfer it to a new
space station, Mir.

AMERICA'S FLYING LABORATORY

NASA's answer to Salyut was Skylab, a single space station
that was visited by three crews of astronauts over two years
following its launch on 14 May 1973. They made the trips in
leftover hardware from the curtailed Apollo Moon programme,
docking with the station in their Command/Service Modules,
lifted aloft by the smaller Saturn IB rockets.

Skylab's own launch nearly ended in disaster because
a protective shield deployed too soon and was torn away,
destroying one of the two main solar panels and snagging
the other. The first crew's launch was delayed while a fix
could be organized, but they then carried out repairs during
a spacewalk. Two more crews gave the US experience of
long-duration spaceflight, during which they carried out
astronomical, medical and other scientific experiments.
The final crew spent 84 days in space.

Plans to prolong Skylab's life by lifting it into a higher
orbit with the Space Shuttle, which was nearing completion,
were abandoned and it was left to make an uncontrolled
re-entry, despite the risks that it might rain chunks of debris
over populated areas. It finally came down on 11 July 1979,
scattering large fragments across a thankfully sparsely
inhabited region of Western Australia.

▼ A Soviet Salyut space station with a Soyuz spacecraft attached.

▲▶ America's Skylab in orbit
(above), showing how one of
the main solar panels is missing,
and the mission patch (right).

However, some fell on Esperance, a small town 580 km
east of Perth. After the re-entry, the San Francisco Examiner
offered a US$10,000 prize for the first piece of Skylab to be
delivered to its offices. Stan Thornton, aged 17, picked up
some chunks from the roof of his home in Esperance and
caught the first flight to San Francisco, where he collected
his prize. Meanwhile, town officials decided to fine NASA for
littering their neighbourhood, sending them a bill for US$400,
which was ignored. Thirty years later, a California radio station
appealed for funds from listeners, and the whip-round raised
the overdue penalty.

▲ Astronaut Jack Lousma takes a shower aboard the Skylab space
station in 1973.

THE MIR SPACE STATION

The Soviet Union ended its Salyut programme to begin building a more ambitious orbiting structure that would consist of several sections, or modules. Mir, which translates to 'community' in this context, began construction in February 1986 and was completed in 1996.

The project, which was intended to see a permanent human presence in low-Earth orbit, saw great political change occur as the Soviet Union collapsed at the end of 1991 and control was assumed by the new Russian Federation. With the change came increased international collaboration and visits by a number of astronauts from the United States and Europe. It lasted as an orbiting residence and workplace for 15 years – three times as long as originally planned.

Mir's core module was the first to be launched, providing basic living quarters, and was followed by another six sections. A year later the Kvant-1 module provided an astronomical observatory, including X-ray and ultraviolet telescopes, a camera and astrophysical experiments.

In 1989, Kvant-2 was launched to add more scientific experimental facilities, the luxury of a shower for the astronauts, and an airlock to allow spacewalks. The following year, the Kristall module was added to allow technological and biological experiments, plus instruments to observe the Earth.

Following Russia's inheritance of the Mir project, its Spektr module was delivered in 1995 to include instruments to study the Earth's atmosphere and environment, to generate power via solar arrays, and to house accommodation and experiments for NASA astronauts. The same year, a new docking module was added to allow the Space Shuttle to visit.

The final module, called Priroda, was another intended for Earth-observation studies, including how the atmosphere interacted with the oceans. Twelve countries contributed experiments, enhancing Mir's now international reputation.

The first visitors to Mir in space were cosmonauts Leonid Kizim and Vladimir Solovyov, who transferred experiments and other instruments from the now redundant Salyut 7. Several other Soyuz flights followed, bringing astronauts from Syria, Afghanistan and France as well as Soviet cosmonauts.

▼ The Space Shuttle *Atlantis* is connected to Russia's Mir space station in this photo from a Soyuz spacecraft in July 1995.

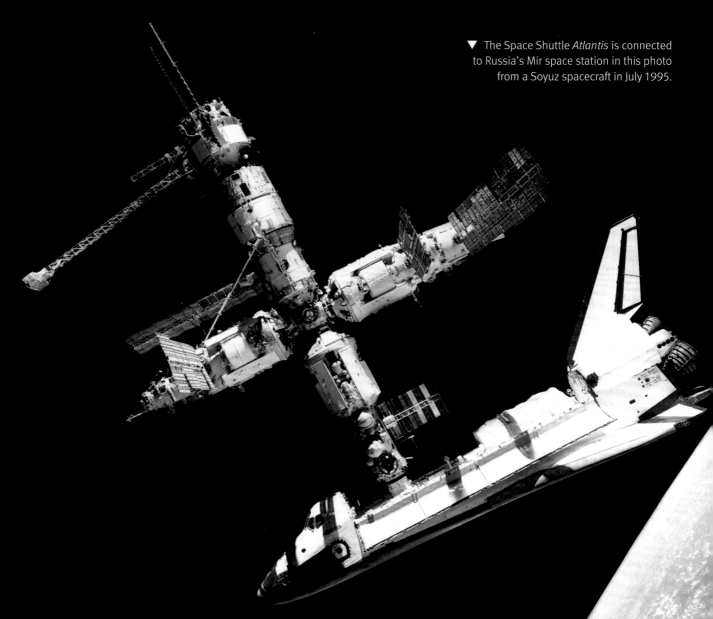

▲ Ten astronauts and cosmonauts crowd the Mir space station following the docking of Space Shuttle *Atlantis* in June 1995.

Following the fall of the Soviet Union, Russia found it lacked funds to complete construction of Mir until the United States stepped in to turn it into a collaborative project. The first US flight to the orbiting complex was by Space Shuttle *Discovery*, which rendezvoused in February 1995 as a dry-run for sister ship *Atlantis* the following June.

Mir survived a serious fire in February 1997 that began in an oxygen generator and burned for several minutes, filling the station with smoke. And in June

the same year, with British-born NASA astronaut Michael Foale aboard, it was rammed by a Progress cargo ship, seriously damaging the Spektr module. As the crew battled to seal off the compartment to stop air escaping, Mir went into an uncontrollable spin. It was only corrected when ground controllers fired the station's engines.

Despite a campaign to extend its life, perhaps in private hands, Mir was brought out of orbit in March 2001, breaking up in a shower of spectacular fireballs over the South Pacific, east of New Zealand.

MIR FACT FILE

- Cosmonaut Valeri Polyakov set the record aboard Mir in 1995 for the longest single stay in space at 437 days, 17 hours and 38 minutes.
- Mir made more than 86,000 orbits of the Earth during its lifetime, travelling at 17,200 km/h.
- The space station was host to 125 astronauts from different nations.
- Visits were made by 31 crewed spacecraft, including nine by the Space Shuttle, plus 64 cargo ships delivering supplies.

LOSS OF FREEDOM

During the 1980s, when the Soviet Union was still in place, the United States had plans to build its own large space station, to be called Freedom. However, the project was abandoned in favour of collaboration on what would become the International Space Station.

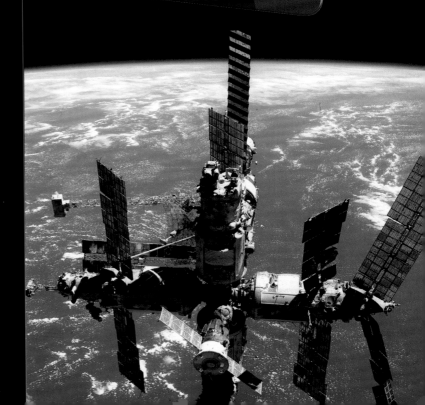

A view of the assembled Mir space station in orbit around the Earth.

THE SPACE SHUTTLE YEARS

A major advance in spaceflight came with the Space Shuttle, an American design that became the world's first reusable crewed vehicle. Though launched in a conventional manner, attached to rocket boosters, the craft would behave like a spaceplane on its return from orbit, landing on a runway just like a conventional aircraft. Once down, it could reasonably quickly be prepared for its next flight. It became a dependable workhorse for NASA for 30 years, with five vehicles in the fleet making regular trips to space, though its success was marred by tragedies that saw two of the spacecraft destroyed with the loss of their crews.

The go-ahead for the Shuttle's development was given by President Nixon in 1972 as the Apollo programme was winding down, but the concept had been around since the 1950s. Designs were drawn up for various spaceplanes, though few had got beyond the drawing board.

Once the project was announced, with its official name the Space Transportation System (STS), there was much competition to build the Shuttle. The final design combined a reusable spacecraft with customary rocket launchers. Two solid rocket boosters would fly attached to a giant external fuel tank. The boosters would fall away, then parachute, soon after launch, to be recovered from the sea, dismantled and reused, but the main tank would get jettisoned above the atmosphere to burn up over the Pacific Ocean.

After typically spending around two weeks in orbit, the Shuttle would fly back to Earth, insulation tiles on its underside protecting it against the fierce heat of re-entry. Then it would glide to a steep landing, at twice the angle of a commercial airliner, on a runway. A parachute would help bring it to a stop, ready to be prepared for its next mission.

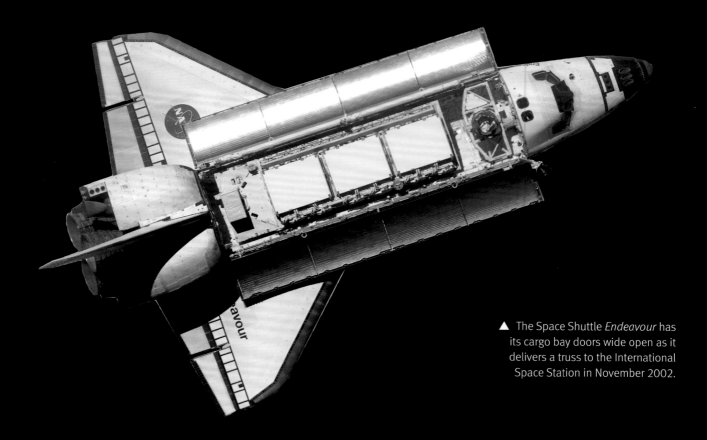

▲ The Space Shuttle *Endeavour* has its cargo bay doors wide open as it delivers a truss to the International Space Station in November 2002.

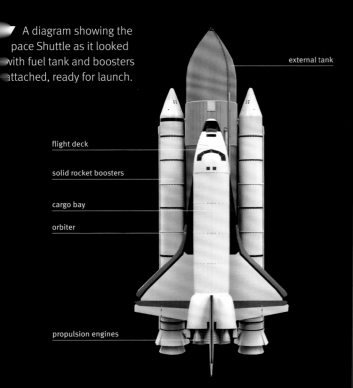
A diagram showing the Space Shuttle as it looked with fuel tank and boosters attached, ready for launch.

external tank

flight deck

solid rocket boosters

cargo bay

orbiter

propulsion engines

Before any Shuttle flew in space, a prototype was built and named *Enterprise*, to the delight of fans of TV's *Star Trek*. Rather than be launched by a rocket, *Enterprise* was carried aloft on top of a NASA Boeing 747 jumbo jet, to be released for flight tests at subsonic speeds in the atmosphere.

Five Shuttles were built for space: *Columbia*, *Challenger*, *Discovery*, *Atlantis* and *Endeavour*. They were named after maritime ships that had made historic voyages of exploration and discovery. Landing sites were at either Kennedy Space Center, Florida, or Edwards Air Force Base, in California, though other strips around the world were available in case of emergency.

The shortest flight was *Columbia*'s second (STS-2) in 1981, lasting 2 days, 6 hours, 13 minutes and 12 seconds. The longest was by the same spacecraft on flight STS-80, at 17 days, 15 hours, 53 minutes and 18 seconds in 1996.

It looks like a Shuttle but this is the Soviet Buran following its return from its only spaceflight in 1988.

SOVIET COPYCAT

America's Soviet rivals developed their own spaceplane, which turned out to bear an uncanny resemblance to the Space Shuttle. Named Buran, meaning 'blizzard', it was designed to be launched by an Energiya rocket, with a central main stage and four attached boosters. After initial tests, the spacecraft made one unmanned spaceflight in 1988, completing two orbits before landing at the Baikonur Cosmodrome. But the programme was cancelled with the collapse of the Soviet Union and the Buran never flew again.

▼ *Columbia* makes the first Space Shuttle launch into space on 12 April 1981, with astronauts John Young and Robert Crippen aboard for a 54-hour mission.

SPACE SHUTTLE FACT FILE

- Shuttle flights carried 306 men and 49 women on 134 flights and 20,952 Earth orbits.
- They flew a total of 864,401,219 km, equivalent to eight return flights to Mars.
- Pioneer astronaut John Glenn became the oldest person to fly in space, aged 77, on Space Shuttle *Discovery* in 1998.
- Each Shuttle included about two and a half million moving parts.

VERSATILITY IN SPACE

The Space Shuttle made its first flight into orbit on 12 April 1981, with a crew of two. *Columbia* was commanded by veteran astronaut John Young, assisted by pilot Bob Crippen on his first spaceflight, and they spent a little over 2 days 6 hours circling the Earth before landing at Edwards Air Force Base, California.

Challenger's first flight came in April 1983, followed by *Discovery* in August 1984 and *Atlantis* in October 1985. Following the loss of *Challenger* in 1986, a fifth Shuttle, *Endeavour*, was later added to the fleet and first flew in May 1992. The Shuttle proved itself to be a versatile vehicle that could carry out a range of operations in orbit around the Earth. As well as simple spaceflights, its vast payload bay was capable of carrying a large amount of cargo into space.

Tasks included deploying satellites directly into orbit, and delivering space telescopes, including Hubble, to which the fleet later returned a number of times to service and repair. Satellites were also repaired in orbit by spacewalking astronauts, or retrieved to be brought back to Earth.

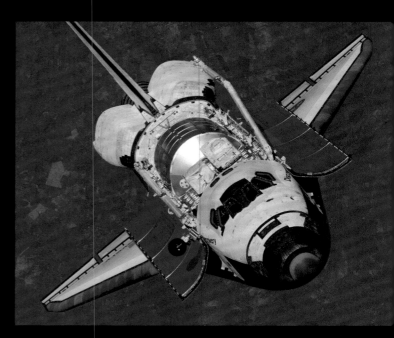

▲ The Space Shuttle *Discovery* approaches the International Space Station carrying a new module in July 2006, during construction of the orbiting outpost.

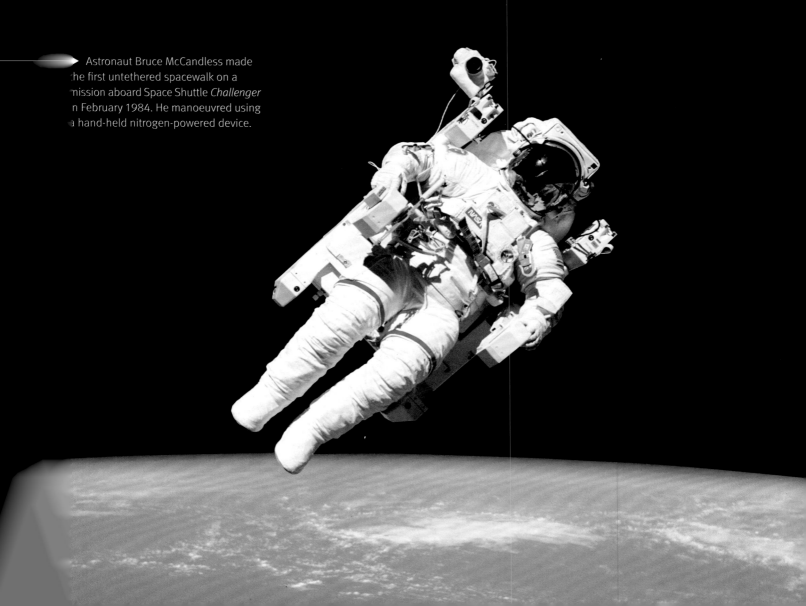

Astronaut Bruce McCandless made the first untethered spacewalk on a mission aboard Space Shuttle *Challenger* in February 1984. He manoeuvred using a hand-held nitrogen-powered device.

Interplanetary probes were launched on their way from the Shuttle, including Magellan to study Venus, Ulysses to orbit and observe the Sun, and Galileo on a mission to rendezvous with Jupiter and orbit the giant planet for several years, studying it and its many moons.

Early Shuttles paid nine visits to the Mir space station, following the new spirit of collaboration between the United States and Russia, following the demise of the Soviet Union in 1991. And then the fleet's final major operation was to assist in the construction of the International Space Station (ISS), carrying several of the larger modules, equipment and other parts, as well as delivering crews to live and work in the orbiting outpost.

The final spaceflight by a Shuttle was made in July 2011 by *Atlantis* to deliver equipment to the ISS. The surviving, decommissioned craft – *Atlantis*, *Discovery* and *Endeavour* – were put on display at museums and visitors' centres across the United States, along with the original test vehicle *Enterprise*, which never flew in space.

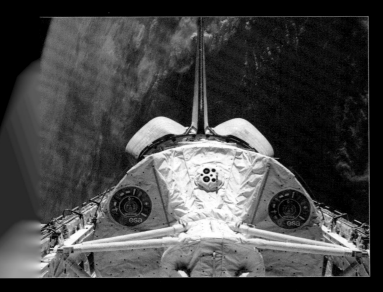

▲ A shot of the European Spacelab, a mini space station, making its maiden flight in the bay of Space Shuttle *Columbia* in 1983.

SPACELAB

An important European contribution to the Shuttle was Spacelab, a small, reusable space station that flew inside the craft's payload bay rather than separately in orbit. Airlocks and a tunnel allowed astronauts to access the laboratory directly from the spacecraft's flight deck.

The flexible arrangement of modules and pallets allowed Spacelab to be used as living quarters as well as a working environment, and the laboratory could be tailored for specific missions, whether to carry out astronomical observations, to study the Earth, or to perform life-science experiments on plants or crew members.

SPACE SHUTTLE – THE STATS

Height of Shuttle stack at launch: 56 m
Length of Shuttle orbiter: 37 m
Height of Shuttle on runway: 17.4 m
Wingspan: 24 m
Orbital velocity: 27,875 km/h

In all, 22 Spacelab missions were flown by the Space Shuttle, and it gave astronauts great experience in working in a shirtsleeve environment in space, helping to prepare them for the upcoming International Space Station. A US version, Spacehab, was also later developed and flown.

▶ Three astronauts on a spacewalk outside the Space Shuttle *Endeavour* catch hold of the giant communications satellite Intelsat VI to replace its engine in 1992.

▼ The Hubble Space Telescope is brought into the bay of *Endeavour* in 1993 for its first maintenance including fitting an optical corrector.

SPACE SHUTTLE TRAGEDIES

CHALLENGER

Just as America was getting used to the success of the Space Shuttle as a regular and reliable vehicle into space, there came tragedy. On 28 January 1986, the orbiter *Challenger* blew up just 73 seconds after the launch of mission STS-51-L, killing its crew of seven astronauts. Spectators at the Kennedy Space Center, and those lining nearby Florida beaches, watched in horror as the explosion left a twisted pattern of smoke plumes, while debris rained down over the region.

The disaster happened after an unusually cold night before launch that had caused a gasket called the O-ring to fail on one of the solid rocket boosters. It allowed burning gas under pressure to reach the external fuel tank.

The crew who perished in the disaster included Christa McAuliffe, who was intended to be the first teacher in space, her adventure followed by children across the United States. The others who died were Commander Dick Scobee, Michael Smith, Ellison Onizuka, Judith Resnik, Ronald McNair and Greg Jarvis.

▲ The *Challenger* STS-51-L crew, pictured clockwise from top left: Ellison Onizuka, Christa McAuliffe, Greg Jarvis, Judith Resnik, Ronald McNair, Dick Scobee and Michael Smith.

The accident caused a crisis at NASA as the Space Shuttle programme was grounded for nearly three years while the cause of the disaster was investigated and new safety measures introduced. It became clear that a senior engineer with contractors responsible for the solid motors had tried to stop the launch because he feared the below-freezing temperatures would affect the rubber seals, but he was allegedly overruled.

▶ The terrible sight seconds after launch of the Space Shuttle *Challenger* as a huge explosion destroys the spacecraft. Seven brave astronauts, pictured above, perished in the tragedy.

The launch came on the day that President Ronald Reagan was due to give his annual State of the Union address. Instead, he paid tribute to the bravery of the astronauts in a moving speech to a shocked world.

COLUMBIA

The second disaster to hit the Space Shuttle programme came not at launch but at mission's end, as seven astronauts were returning from 16 days in orbit, aboard *Columbia*. Everything had gone well in space and the crew of five men and two women had performed around 80 experiments as planned.

On 1 February 2003, the spacecraft dipped to re-enter the atmosphere for a landing in Florida to end its 28th mission, but then things went terribly wrong. Unbeknown to the astronauts on board, their spacecraft's left wing had been damaged during launch when a chunk of insulating foam on the external fuel tank fell and hit its leading edge. Heat-resistant tiles were cracked and, in the searing heat of re-entry, allowed hot gases to penetrate and the spacecraft to break up. As NASA mission controllers grew concerned at the lengthy silence from the spacecraft, others were observing

the terrible sight of multiple streaks in the sky as fragments of the doomed Shuttle fell over Texas and Louisiana.

The victims who perished were Commander Rick Husband, and his crew, Willie McCool, Michael P. Anderson, Israeli astronaut Ilan Ramon, Indian-born Kalpana Chawla, David Brown and Laurel Clark.

This time, President George W. Bush addressed the nation to praise the astronauts' courage. Memorial services were held at the Johnson Space Center, Houston, which is home to NASA's astronaut corps, and Washington National Cathedral.

Following the tragedy, the Space Shuttle was again grounded for three years while an investigation was mounted and NASA again attempted to learn from it and make future spaceflights safer. The Shuttle returned to flight with a *Discovery* mission in July 2006.

All subsequent missions, apart from one to service the Hubble Space Telescope, were made to the International Space Station, which could be used to protect astronauts in the event that their spacecraft had been similarly damaged. Small amounts of insulating foam were seen to fall from the external tank on various missions, but never again did they have such terrible consequences.

▲ The tragic crew of *Columbia* mission STS-107 pictured weightless in space before their fateful return – (clockwise from left) Kalpana Chawla, David Brown, Willie McCool, Michael Anderson, Ilan Ramon, Laurel Clark and Rick Husband.

◄ Launch of Space Shuttle *Columbia* on mission STS-107 which was to end in tragedy. The spacecraft broke up on re-entry over the United States.

THE INTERNATIONAL SPACE STATION

The fall of the Soviet Union in 1991, and better relations between East and West, led the United States to drop plans for a new space station of its own and instead to embark on a project that would see several nations collaborate in space. Thus the concept for the International Space Station (ISS) was born.

▲ The International Space Station speeds in its orbit above a cloud-shrouded blue Earth.

Rivalry that had successfully driven the start of the Space Age was replaced by a new spirit of co-operation. Fifteen governments signed up to be part of the new adventure, with NASA and the space agencies of Europe, Canada, Russia and Japan working together in building it.

Construction began in November 1998 when Russia launched the Zarya module, which had originally been designed for Mir. Since then, the ISS has grown to include 15 pressurized modules, contributed by different nations, with one more laboratory intended to be sent by Russia

in 2017. The hardware also includes airlocks, trusses, vast solar arrays to provide power, and robotic arms to manipulate visiting craft and other items.

The combined sections give astronauts a comfortable environment in which to live and work, inside the largest structure ever built in space.

The ISS has been continuously occupied since November 2000 and has hosted astronauts from around the world. They live and work in weightlessness while speeding round the Earth at 8 km/sec, about 355 km above the ground.

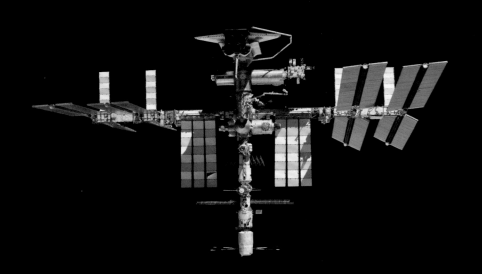

◀ Viewed from a Soyuz in 2011, the ISS is seen with various spacecraft attached, including the Space Shuttle *Endeavour* at top, and ESA's ATV *Johannes Kepler* below.

▶ The ISS has been constructed from 15 modules, plus other assorted hardware.

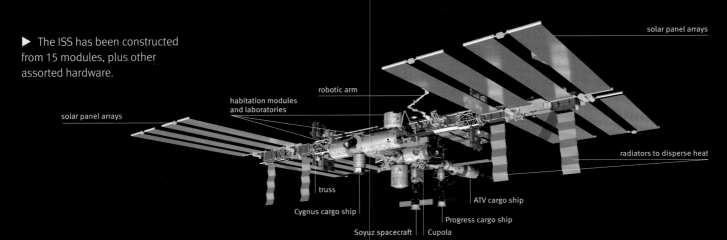

solar panel arrays

robotic arm

habitation modules
and laboratories

solar panel arrays

truss

Cygnus cargo ship

Soyuz spacecraft   Cupola

Progress cargo ship

ATV cargo ship

radiators to disperse heat

Since the Space Shuttle was retired, all astronauts have had to be transported to and from the ISS by Russian Soyuz spacecraft, but cargoes have been delivered by Russian Progress freighters, Europe's Automated Transfer Vehicle (ATV), Japan's H-II Transfer Vehicle (HTV) and two commercial American spacecraft, Dragon and Cygnus. As well as delivering supplies, these spacecraft are also used to help boost the ISS to a higher orbit as it gradually sinks due to slight atmospheric drag.

Astronauts are kept very busy carrying out a vast range of scientific research, taking advantage of the weightless environment to conduct experiments to advance medical, biological and physical knowledge, as well as studying the Earth and space, and developing new technology. They are also helping to prepare humans for long-duration missions, including a return to the Moon and missions to asteroids and Mars. In March 2016, NASA astronaut Scott Kelly and Russian cosmonaut Mikhail Kornienko returned to Earth after a record single stay of 340 days on the ISS.

OBSERVING THE ISS

With its highly reflective solar panels that are half the size of a football pitch, the ISS is a very bright object when it appears in the night sky. It shines brighter than any star and can be seen to move rapidly, rising from the western horizon and disappearing in the east. Its steady glow will help distinguish it from aircraft with their flashing lights.

The orbit of the ISS is inclined, carrying it from latitude 51.6°N to 51.6°S on every revolution. Its track over the Earth varies over time, so that occasions to see the ISS during night hours come in sequences for a particular location. On late evenings, the ISS will be seen to fade quickly and disappear as it enters the shadow of the Earth in space – it only shines when it is catching the light from the Sun.

The ISS is easy to photograph if you can set a time exposure with your camera on a tripod. It will then appear as a trail across the sky.

THE ISS FACT FILE

- The ISS has as much room inside as in a six-bedroomed house or a Boeing 747 jumbo jet.
- It is four times the size of Russian forerunner Mir.
- Its solar panels cover 2,500 sq m, providing enough electricity for 40 homes.
- The electrical wiring in this orbiting home is 13 km long!
- By the end of 2015, ISS astronauts had carried out 192 spacewalks.
- In that time, it had been visited by 222 people from 18 countries.

▶ NASA astronaut Barry Wilmore is pictured carrying out maintenance duties outside the International Space Station on a spacewalk in February 2015.

WORKING ON THE ISS

The International Space Station typically hosts a crew of six. Each astronaut stays for about six months, and they arrive three at a time every three months to take over from the three who are leaving. Each new team of six astronauts aboard is assigned an Expedition number. They are kept very busy on the ISS because time in space is precious, and they occupy themselves either with experiments or with maintenance tasks.

The ISS is above all an orbiting laboratory intended to carry out research in a unique environment. Some of the experiments are tests on the astronauts themselves and their health to see how they adapt to being in space, and others might be on plants and small organisms. Other data is gathered on Earth science by observing such things as glaciers, agricultural areas and coral reefs.

Technologies can also be tested in the microgravity environment to see how they will perform on long-duration space missions. And physical science research discovers things such as how crystals grow, and how fire and fluids behave in such an environment. The ISS is also a valuable classroom where schools can get involved with the research to teach students science and engineering.

Apart from the science, there are routine maintenance tasks that have to be performed to keep the ISS functioning. Support systems need to be checked constantly, computer equipment must be updated, and filters and other items

▲ NASA astronaut Tim Kopra sets up an experiment to find out how fabrics burn in space.

cleaned. Mission controllers monitor the status of equipment on the space station and send the astronauts a list of tasks to carry out each day.

When a cargo ship arrives with fresh supplies or new equipment, this is a busy time for the crew who have to unload it and then fill the empty capsule with unwanted items, just as you might put out rubbish for your refuse collector, so that it can be disposed of by burning it up in the atmosphere.

Apart from such routine housekeeping, there are a few occasions when something needs to be attended to outside the Space Station, such as some new wiring, or a repair or maintenance to some external equipment. At such times, a couple of astronauts will usually suit up to venture outside for a spacewalk to carry out the task. Such EVAs can last six or more hours at a time before they return inside.

▼ NASA astronaut Robert Curbeam and ESA crewmate Christer Fuglesang work on construction of the International Space Station in 2006, high above New Zealand.

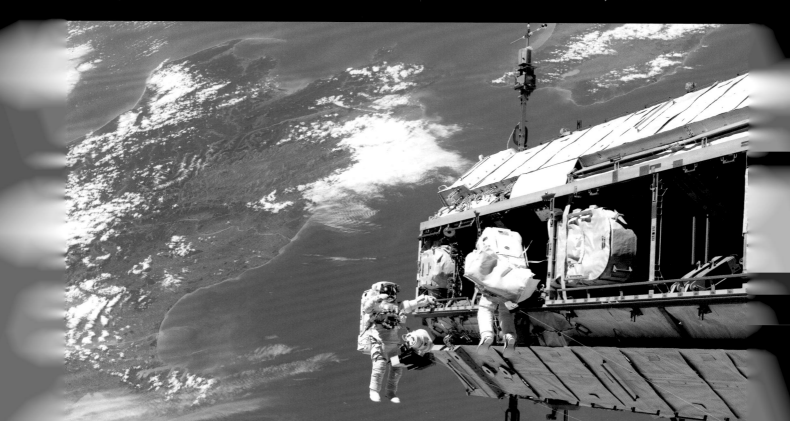

THE SPACESUIT

The spacesuit has evolved considerably since early spaceflight. Today's US suits are assembled from several ready-made components, to fit the individual astronaut, and are made from many layers of different materials including fabric, mylar, nylon, spandex, stainless steel and other high-strength composite materials. They are white to reflect the sunlight.

The suits have to keep the astronaut's body temperature comfortable in the hostile space environment, protect against micrometeoroid impacts, but be flexible enough to allow the astronaut to work. An electrical harness provides communications and also monitors the astronaut's vital functions such as breathing, heart rate and temperature. An undergarment is laced with plastic tubing for cooling and ventilation. A life-support system provides oxygen, power and cooling water, and removes exhaled carbon dioxide. It also holds radio equipment and a warning system. There is a secondary oxygen pack for emergencies.

▲ NASA astronaut Karen Nyberg prepares to capture a Japanese spacecraft, the Kounotori HTV-4, while working in the ISS Cupola in 2013.

The astronaut drinks from a bag of water held within the suit. Gloves are designed to allow astronauts to use tools, and the helmet includes a visor for EVAs to protect against bright sunlight. Loops on the suit allow the astronauts to tether themselves to the hull of the Space Station for safety while working outside. Russian suits have similar functions.

Inside the ISS, astronauts are able to wear casual gear, usually their flight suits, but change into pyjamas when they are ready for bed. Shuttle astronauts wore a pumpkin-coloured suit for take-off and landing.

▲ Russian cosmonaut Oleg Kotov performs maintenance outside the ISS in 2010.

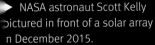

▶ NASA astronaut Scott Kelly pictured in front of a solar array in December 2015.

LIVING ON THE ISS

RELAXING IN SPACE

Though astronauts are kept busy, they are also given some downtime each day when they can take a break and relax. During these breaks they might watch movies, read books, play games, answer emails or phone their families over the internet. Canadian astronaut Commander Chris Hadfield famously took his guitar with him to play David Bowie's hit 'Space Oddity'. The astronauts also enjoy spending time in the Cupola, a viewing platform attached to the ISS, where they can simply enjoy the view down on to the Earth, and maybe take some photos.

SPACE FOOD AND DRINK

Unlike the earliest astronauts who survived on unappetising tubes of paste or bite-sized cubes, today's astronauts are given a more varied and attractive selection of foods, choosing their menus from an extensive list months before they even fly. Prepared by expert nutritionists, they come in disposable packages, often frozen, to be heated in the ISS galley.

▼ NASA astronaut Tracy Caldwell Dyson takes time to relax and enjoy the view over the Earth from the ISS Cupola in September 2010.

The astronauts eat three meals a day, with sauces to liven them up. Salt and pepper are in liquid form as grains could float away and be inhaled or enter the eyes. The visit of an unmanned cargo ship to the Space Station is always welcomed as it usually brings a selection of fresh fruit and other treats.

Water is a very precious commodity and extremely expensive to bring from Earth. Though some is delivered in the cargo ships, much of it is reclaimed and recycled via two different water-recovery systems on the ISS, in the Russian and American sections. They are able to process waste water from showers, basins, other crew systems and vapour condensed from the air.

EXERCISE IN SPACE

Weightlessness is known to have a negative effect on the human body, so astronauts spend a couple of hours a day doing exercises to try to counter bone and muscle loss. Special equipment is designed to give them a healthy workout, including a treadmill so that they can go-jogging.

GOING TO THE BATHROOM IN SPACE

Astronauts cannot use a conventional shower on the Space Station, Instead they squirt warm water into towels so that they can wipe themselves down, rather like having a sponge

▲ British astronaut Tim Peake juggles with some fresh fruit in the weightlessness of space.

TIME FOR BED

The ISS is orbiting so quickly that there is no day and night pattern to match that on the ground. But like any-one else, astronauts need their rest, so they are allocated sleep periods with a similar cycle to home. There are no beds in the traditional sense – instead the astronaut wraps up in a sleeping bag, which is attached to the wall of a small crew cabin so that he or she does not literally drift off to sleep and collide with something!

For solid waste, the astronaut takes a seated position, with restraints to help stay in place, and again a fan helps the process along, rather like a vacuum cleaner. The waste is collected each time in a bag, which is sealed and stored for later disposal.

bath in hospital. They wash their hair using water and a special no-rinse shampoo before towelling the hair dry. They can also brush their teeth and shave, just as on Earth.

It is natural to wonder how humans perform their bodily functions in space. While astronauts still have a kind of nappy in their spacesuits for EVAs and use during launch and landing, the ISS is equipped with two special lavatories to make life more comfortable. To urinate, the astronaut attaches a length of tubing with a funnel, taking care to turn on a fan first to extract the liquid. The urine is not wasted but converted into fresh water for drinking.

▲ Canadian astronaut Chris Hadfield relaxes by playing the guitar in the Cupola. His version of David Bowie's 'Space Oddity' was an internet hit.

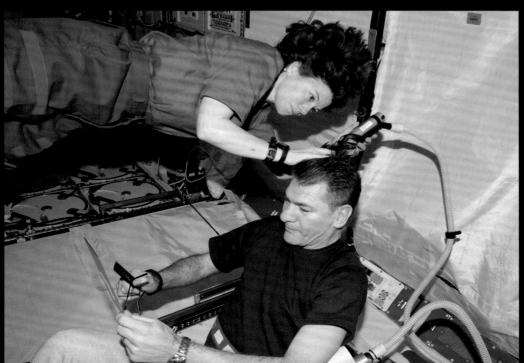

◀ NASA astronaut Cady Coleman gives Italian colleague Paolo Nespoli's hair a quick trim in the Kibo laboratory on the ISS in January 2011. A vacuum cleaner is attached to the clippers to catch stray hair.

HEALTH IN SPACE

Astronauts are a special breed who undergo extensive training before they ever fly into space. However, they are not superhuman. Frequently they experience space sickness when they first arrive in orbit, and sometimes on their return to Earth too. Officially known as 'space adaptation syndrome', it causes the astronaut to feel dizziness and disorientation, with symptoms ranging from mild headaches to nausea. Thankfully it is a malaise that usually wears off after two or three days as they become adapted to their new environment.

Medical scientists believe the cause is similar to what can make people feel travel sick when in a car and trying to read a book. There is a mismatch between the inner ear that detects the car's motion, and the eyes that are staring at a set of fixed words. It is hard to predict which astronauts will get ill, and has nothing to do with how fit they are. Experienced jet pilots or gymnasts used to their bodies moving in extreme ways can suffer the worst when it comes to spending time in orbit.

Though space sickness is annoying, more important health problems occur over extended stays. Astronauts on the ISS have been found to suffer bone and muscle wastage in the weightless environment. It is why they spend a lot of time every day exercising to try to compensate for the damage being done to their bodies – and it's also why they take some time to recover on their return to Earth. Images of astronauts landing in Soyuz spacecraft always show them being carried from their re-entry capsule in chairs. They would probably lose consciousness if they tried to stand before adapting to gravity again.

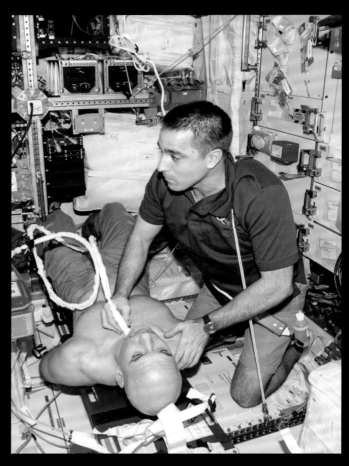

▲ NASA astronaut Chris Cassidy performs an ultrasound experiment on ESA's Luca Parmitano in 2013 to investigate the effects of a prolonged stay in space on the spine.

More serious threats to health come on trips away from Earth. A violent solar flare during a mission to the Moon or Mars could bring a real risk of radiation sickness for the voyagers. Eruptions occurred from a huge sunspot in 1972, thankfully during the gap between the Apollo 16 and Apollo 17 missions. Any Moonwalker at the time could have received dangerous levels of radiation.

Astronauts on the International Space Station are already better protected than those on long-distance missions. The ISS is heavily shielded in certain modules but it also orbits inside Earth's magnetic field, which adds a good amount of protection. Interplanetary missions will require special shielding, and research has begun into creating a local artificial magnetosphere for spacecraft.

◀ NASA's Karen Nyberg conducts an ocular health exam on herself aboard the ISS to check how being in space affects pressures in the eye.

◀ Russian cosmonauts Mikhail Kornienko (left) and Sergey Volkov (centre) together with NASA astronaut Scott Kelly, recover in chairs following the return of the Soyuz spacecraft from the ISS in March 2016.

▼ Belgian astronaut Frank De Winne keeps fit by running on a treadmill in the ISS's Harmony module.

Long periods in space have also been found to affect astronauts' eyesight. The effects have been observed in some returning from six-month missions and are due to body fluids producing greater pressure around the brain and at the back of the eye, causing damage to the optic nerve.

After a year in space, the body's immune system has been found to become weaker at the same time as microbes and germs mutate in the stomach. Serious illness could occur from the exchange of bugs between members of a crew.

Apart from the physical risks, researchers are also investigating the mental pressures on astronauts undergoing a lengthy space mission. Feelings can range from simple boredom and isolation to disagreements with other crew members. Some cosmonauts are said to have refused to speak to each other for months. Such a lack of communication could imperil a long-duration mission.

The worst nightmare comes if an astronaut suddenly requires emergency surgery. The situation is not so bad on the ISS because the patient could be flown home relatively quickly using one of the spacecraft permanently attached to the orbiting outpost. But it will be very different in deep space, far from a hospital, such as on an expedition to Mars. For this reason, American scientists are developing a miniature robot that could perform keyhole surgery in space.

CHINA ENTERS THE SPACE AGE

China, one of the great world powers, came late to the space exploration party, but has been rapidly making up for lost time. Though crewed spacecraft were being designed in the 1960s, the first satellite was not launched until 24 April 1970, from the Jiuquan Satellite Launch Centre in the Gobi Desert. Since then, China has followed a largely independent space programme, although it is a signatory to most international outer space treaties and is working with a number of respected space scientists around the world. However, it is not a partner in the ISS and NASA is barred from working with the China National Space Administration.

CHINESE CREWED MISSIONS

After a number of false starts, China finally focused on human spaceflight at the tail end of the last millennium. After four unmanned test flights, Shenzhou 5, launched by a Long March 2F rocket, was the first Chinese mission to put a human in space on 15 October 2003, becoming the third nation to do so. Fighter pilot Yang Liwei spent 21 hours making 14 orbits before landing in the grasslands of Inner Mongolia. Chinese astronauts are known as taikonauts.

The Shenzhou spacecraft bears some resemblance to Russia's Soyuz, though it is larger. It is a Chinese design but incorporates some bought-in Russian technology, and combines an orbital module in which the crew can operate, a re-entry module and a service module providing power and life-support systems.

Two years later, on 12 October 2005, Shenzhou 6 carried a crew of two into space. Fei Junlong and Nie Haisheng made 75 orbits of the Earth before landing more than four days later.

The next mission, Shenzhou 7, launched on 25 September 2008 and included China's first EVA, or spacewalk. Zhai Zhigang and Liu Boming spent 20 minutes outside, while the third crew member, Jing Haipeng, remained in the spacecraft.

Shenzhou 8 made an uncrewed flight into orbit in November 2011 to dock with China's first space station, Tiangong-1, a simple single module with two large solar panels.

Liu Yang became China's first woman in space on Shenzhou 9, with crewmates Jing Haipeng and Liu Wang. The spacecraft, which launched on 16 June 2012, also docked with Tiangong-1. A year later, on 11 June 2013, Shenzhou 10 carried another female taikonaut, Wang Yaping, flying with

◀ A Long March rocket carrying the unmanned spacecraft Shenzhou-8 blasts off from the Jiuquan Satellite Launch Centre in north-west China in November 2011.

▲ China's first woman in space, Liu Yang (left) prepares for her flight on Shenzhou 9 with fellow taikonauts Jing Haipeng and Liu Wang at the Jiuquan Satellite Launch Centre in June 2012.

male colleagues Nie Haisheng and Zhang Xiaoguang to dock with the space station.

The Shenzhou programme is scheduled to continue with future spacecraft visiting a new and larger space laboratory, Tiangong-2, and a third, modular orbiting space station is also planned.

Having already sent probes to the Moon, including the soft-lander Chang'e 3 with its rover Yutu in 2013, China seems focused on sending taikonauts to the Moon to set up a base, and later to Mars.

▲ Taikonaut Zhai Zhigang waves from the Shenzhou 7 spacecraft as he makes his spacewalk in September 2008.

SPACE JUNK

A growing hazard for astronauts, particularly aboard the International Space Station, has been the proliferation of space junk – from abandoned rocket stages, and items lost by spacewalking astronauts, to the hundreds of thousands of small fragments from defunct satellites.

Working satellites, including the ISS, have to make several manoeuvres each year to avoid the larger pieces that space agencies such as NASA and ESA are able to track. Even a small fragment, orbiting the Earth at many times the speed of a rifle bullet, could cause huge damage if it hit a satellite's electronics or pierced the hull of the Space Station. NASA has a number of times ordered the crew of the ISS to shelter in a Russian Soyuz spacecraft, which always stays attached as an available 'lifeboat', until the danger has passed.

The hazard increased in 2007 when China performed a 'star wars' test by destroying one of its old weather satellites with a missile, creating a cloud of debris. More space junk was created when two satellites collided in 2009. The space agencies are taking the problem very seriously and scientists are designing space-craft that could sweep clean the space lanes by collecting and dis-posing of much of the orbiting litter.

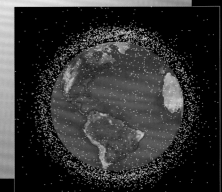

EXPLORING THE OUTER SOLAR SYSTEM

Jupiter's moon Europa emits plumes of water vapour as it hangs in front of the Solar System's biggest planet, with a raging storm dubbed the Great Red Spot in view. Mighty Jupiter is the first of four giant outer worlds made up mainly of gas and ice. Space probes have transformed our knowledge of this region of the Solar System and are beginning to inform us about dwarf planet Pluto and other smaller bodies beyond.

THE GRAND TOURS

With spacecraft already having visited the Moon and nearby planets, NASA turned its attention towards exploration of the outer Solar System. It was realized that a rare line-up of the planets in the late 1970s would allow more than one planet to be encountered by a single probe. The ultimate aim was for a Grand Tour to be made by two Voyager craft, but before they were sent, NASA launched two other spacecraft, Pioneer 10 and Pioneer 11, the latest in a long-running Pioneer programme of planetary missions.

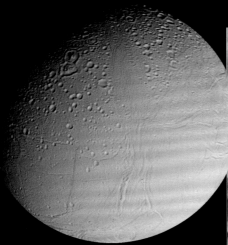

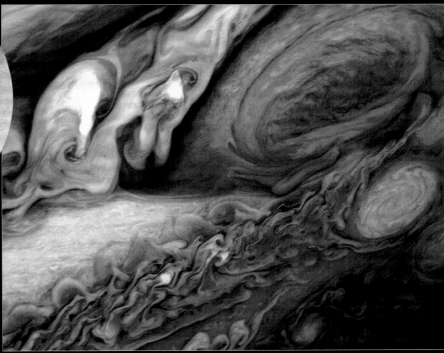

▲ Voyager 2 found that Saturn's moon Enceladus has an icy surface, plus craters and linear cracks.

➤ Jupiter's Great Red Spot and cloud tops resemble an abstract painting in this image from Voyager 1.

PIONEERS 10 AND 11

Pioneer 10 launched in March 1972 from Cape Canaveral, Florida, atop an Atlas–Centaur rocket, and became the fastest vehicle ever at the time, racing past the Moon after just 11 hours. It also became the first spacecraft to fly through the asteroid belt, entering this potentially hazardous zone of rocky fragments in July 1972 and emerging in February 1973, *en route* to the gas giants Jupiter and Saturn.

As Pioneer 10 approached its main target Jupiter, the largest planet in the Solar System, it began to take hundreds of photos of much higher quality than could be managed from Earth. Closest approach to the giant planet was made on 4 December 1973, at a distance of 130,354 km, by which time the probe was travelling at 132,000 km/h. During the flyby, Pioneer 10 also measured Jupiter's magnetic field and got close-up images of three of its main moons – Callisto, Ganymede and Europa.

Sister spacecraft Pioneer 11 launched in April 1973 from Florida and again flew without incident through the asteroid belt, helping allay scientists' fears over how hazardous this region of the Solar System might be. Pioneer 11 flew much closer to Jupiter than Pioneer 10 had, passing just 42,760 km over its belts of colourful cloud on 3 December 1974. But its greater speed of 171,000 km/h meant its instruments were less affected by the planet's powerful radiation.

◀ An artist imagines how one of the twin Pioneer probes might look in the outer reaches of the Solar System.

➤ Saturn's spectacular ring system was revealed to be intricately detailed when Voyager 2 flew past in 1981.

The Jupiter encounter was not the end of Pioneer 11's mission. It used the planet's orbital momentum to boost it onwards towards its second target, Saturn, a journey that would take nearly five years. The spacecraft confirmed that the ringed planet has a magnetic field, first feeling its effects while still 1.5 million km distant. Closest approach was made on 1 September 1979, at a distance of 20,900 km.

Contact with Pioneer 11 was lost in September 1995 when it antenna turned away from Earth. Pioneer 10's last communicatio came in January 2003 before its power source failed. Today, the two Pioneers are heading in different directions out of the Solar System. Pioneer 10 is racing towards the bright star Aldebaran in the constellation of Taurus, while Pioneer 11 heads for the centre of our Milky Way Galaxy in Sagittarius.

LONG-DISTANCE RUNNERS

VOYAGERS 1 AND 2

The Pioneers produced valuable planetary science but they were really the warm-up acts for two other twin probes, Voyager 1 and Voyager 2. Voyager 2 was launched first, in August 1977, from Cape Canaveral on a Titan–Centaur rocket. Voyager 1 lifted off 16 days later in September 1977, taking a faster route that saw it leave the asteroid belt before its sister ship. Voyager 1 reached Jupiter's system of satellites in February 1979, and its cameras, vastly superior to those on the Pioneers, sent back spectacular pictures of moons Amalthea, Io, Europa, Ganymede and Callisto, and discovered that Io has very active volcanoes. It also imaged greater activity than had previously been seen in the planet's swirling clouds, discovered that Jupiter has its own ring, though a pale relative of Saturn's impressive display, and found two new moons, Thebe and Metis. Closest approach to Jupiter was made on 5 March 1979 at a distance of 280,000 km.

Voyager 1 travelled on to become the second spacecraft to visit Saturn, which it got closest to on 12 November 1980. Major findings were an atmosphere around largest moon Titan, a new ring, and four new moons that were influencing,

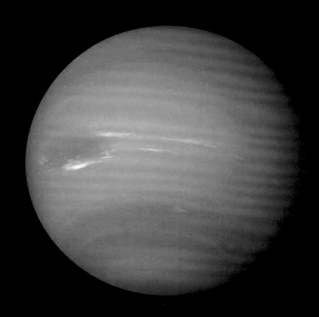

▲ Voyager 2 captured this close-up view of Neptune in August 1989, revealing a great dark spot and neighbouring bright smudge, plus fast-moving features in the clouds.

or 'shepherding', the shape of two other rings. Detailed photos were sent back of the intricate ring system, plus moons Mimas, Enceladus, Tethys, Dione and Rhea.

After leaving Saturn, Voyager 1 turned its cameras backwards to take a mosaic of the planets, including an iconic image of the Earth as a 'pale blue dot'.

When Voyager 2 followed on, it gave further fine pictures of Jupiter and its moons, passing closest to the planet on 9 July 1979. It confirmed how constant volcanic activity is reshaping Io's surface, imaged cracks in Europa's icy crust and discovered a 14th moon.

At Saturn, which it passed closest to on 26 August 1981, Voyager 2 recorded much more detail in the rings. But it then flew on to visit two further planets that had never been visited before or since – the ice giants Uranus and Neptune. It flew by Uranus on 24 January 1986, discovering ten new moons and two new rings.

Uranus's gravity helped sling Voyager 2 on towards Neptune. It took many years to make the journey, finally flying by the eighth planet on 25 August 1989. Discoveries at Neptune included five moons, four rings and a temporary large dark spot in Neptune's cloud tops. A volcano of nitrogen ice was detected on the largest moon, Triton.

The Voyagers each took around 18,000 photos of the Jupiter system and 16,000 of Saturn's retinue. Voyager 2 took a further 8,000 images at Uranus and 10,000 at Neptune.

Voyager 1 was declared to have become the first spacecraft to leave the Solar System and enter interstellar space in August 2012, heading in the direction of constellation Camelopardalis. Voyager 2 was expected to cross the boundary in 2016, heading in the general direction of Sirius, the brightest star in the night sky.

▲ An image of straw-coloured Saturn with its magnificent rings from Voyager 2 in August 1981. Also visible are the moons Tethys, Dione and Rhea, plus the shadow of Tethys appears as a dark spot in Saturn's southern hemisphere.

INTERSTELLAR GREETINGS

The first four probes to be sent heading out of the Solar System were fitted with items that would assist any intelligent aliens who might retrieve them some time in the distant future. The Pioneers each carried a plaque showing the naked figures of a man and a woman, the spacecraft, and diagrams to represent a hydrogen molecule, the layout of the Solar System, and the position of our Sun in relation to the Galaxy and 14 known cosmic beacons called pulsars.

The Voyager probes were each equipped with a Golden Record – a phonographic disk that holds images and sounds from Earth. Sounds include music, greetings in various languages, waves, wind, thunder, birds and whales. Pictures include humans, animals, insects, plants and architecture. Helpfully, a stylus was included in case the aliens have no compatible equipment on which to play the record!

▶ The Golden Record that each Voyager carried to tell any aliens they might encounter in the future about the Earth.

▼ An artist's impression of a Voyager probe flying through deep space.

VOYAGERS 1 & 2 FACT FILE

- The arrangement of the outer planets allowing the Voyager probes to visit them on their Grand Tours occurs approximately every 175 years.
- The speed boost that flybys of Jupiter and Saturn gave to Voyager 2 cut the flight time to Neptune from 30 years to 12 years.
- Though built to last only five years, the two Voyager probes are still communicating with Earth nearly 40 years after they were first launched.

JUPITER – THE GAS GIANT

Jupiter, the fifth planet from the Sun, is also the largest, being big enough to contain more than 1,300 Earths, and more than two and a half times as massive as all the other planets put together. Named after the supreme Roman god, it is the first of the gas giant planets, so called because they are mainly balls of gas rather than rocky like the inner terrestrial planets. However, it is believed that there is a rocky inner core to the planet. Space probes are helping scientists understand Jupiter better.

Despite its incredible size, which is much greater than all the other planets in the Solar System combined, Jupiter rotates rapidly on its axis. A day on Jupiter lasts a little under ten hours, and this speed causes the planet to bulge, making its polar diameter significantly less than at the equator. Astronomers call this shape an oblate spheroid.

▶ This detailed picture of Jupiter was produced from 27 images captured by NASA's Cassini–Huygens spacecraft as it flew past in December 2000, on its way to explore Saturn.

▶ A close-up of Jupiter's Great Red Spot, a huge anticyclone, plus a neighbouring white spot and other cloud features in a picture taken by Voyager 1 in 1979.

The invention of the telescope allowed early astronomers to see Jupiter's disk, and they observed that there were bands of different hues across it. These are the cloud belts, which vary in individual intensity over time, but retain their distinctiveness as different weather zones. Particularly famous in the cloud belts is a feature called the Great Red Spot.

Also revealed in the earliest telescopes were four points of light to either side of the planet, strung out like bright beads, whose positions changed quite rapidly. These are Jupiter's biggest moons – Io, Europa, Ganymede and Callisto – and they are known as the Galilean satellites after the famous Italian astronomer Galileo Galilei who drew attention to them in the early 17th century, having first mistaken them for stars. The discovery of these moons was important because it confirmed that celestial bodies did not all orbit the Earth.

GREAT RED SPOT

Jupiter's best known feature, the Great Red Spot, is a huge and powerful storm that has been recorded by astronomers since at least 1878, and is possibly the resurgence of a phenomenon noticed by the observer Cassini in 1665. Since the last century, it has been steadily shrinking and it is now half the size it once was, though its current diameter of about 16,500 km means it could still easily contain the Earth. Winds at the edge of this anticyclone are blowing at around 360 km/h.

OBSERVING JUPITER

Jupiter is so bright that it resembles a brilliant star in the night sky, though like all the planets it is shining purely by reflecting sunlight. Of the planets, only Venus get brighter, and only Mars at a close opposition can ever briefly become as bright in a midnight sky.

As well as being so luminous, Jupiter gives itself away by shining steadily and not twinkling like bright stars. That is because it is close enough to show as a disk, even though the disk itself is not visible to the eye without optical aid.

Binoculars are enough to reveal those bright Galilean moons. Hold them steady, perhaps on a tripod or by leaning against a wall, and you will see the moons as starlike points.

JUPITER FACT FILE

- Gravity at Jupiter's visible surface is about 2.4 times stronger than on Earth.
- Radiation levels around Jupiter are 1,000 times the safe level for humans.
- Jupiter's polar diameter is 9,300 km less than at the equator and a day at the poles lasts up to five minutes longer.
- Unlike some other planets, Jupiter does not experience any seasons because its polar axis is tilted by only 3.13 degrees to the plane of its orbit.

Their motions will become obvious over just a few hours. You will see them switch from one side of Jupiter to the other, and you will see fewer than four if one or more is in front of, or behind, Jupiter's disk.

A small telescope will make it easier to view the moons, and will allow you to see the belts in the clouds. Don't expect to see them easily straight away. Spend a few minutes watching the yellowish disk and your eye will begin to perceive the subtle differences in shading. The Great Red Spot has not been as prominent in recent years as it used to be, but should be visible unless it is on the far side of the planet when you observe.

▲ A simulation of how Jupiter and its four Galilean moons appear through a small telescope.

DELVING INTO JUPITER

The different bands visible on Jupiter's disk are known as belts or zones and are given individual labels, such as the North Tropical Zone and the South Equatorial Belt. The zones are bright and are where Jupiter's weather systems move from west to east. Dark belts have their prevailing winds blowing in the opposite direction, from east to west. Early observers' sketches of detail in these bands revealed the rapid rate at which the giant planet rotates.

As well as the waves and swirls seen in the bands, and the famous Great Red Spot, occasional white or brown ovals also appear. The swirling material that produces Jupiter's weather includes methane, ammonia and water. Infrared studies with the Hubble Space Telescope have revealed that storms are driven by hot jets of ammonia from within the planet. Jupiter is made up mainly of hydrogen and helium, like a star. It radiates twice as much heat as it receives from the Sun, and if it had been around 80 times more massive it might have become a star itself!

The outer layers of Jupiter's atmosphere are only about 50 km thick and driven by convection currents. At the cloud tops it is a cool −110°C, but the temperatures begin to soar further down as the gases become more compressed, like liquid. At a depth of 15,000 km, they reach more than 5,000°C and the gas behaves more like molten metal. Jupiter's heat energy is generated by the planet collapsing in on itself. At its heart, there is thought to be a solid rocky core about 20,000 km wide.

Jupiter is surrounded by a vast magnetic field – or magnetosphere – that extends beyond its entire system of moons and produces aurorae similar to the northern and southern lights on Earth. It also produces natural radio signals.

Despite its huge size, Jupiter's mass is equivalent only to 318 Earths because it is considerably less dense than a rocky world. The planet is believed to have changed little since the formation of the Solar System.

ULYSSES

The Ulysses spacecraft, launched from the Space Shuttle *Discovery* in 1990 to explore the Sun's effects on the Solar System, was routed via Jupiter. In its flybys the joint NASA/ESA probe detected radiation bursts from the planet, caused by interaction of the solar wind with its magnetic field, plus streams of dust originating from volcanic moon Io.

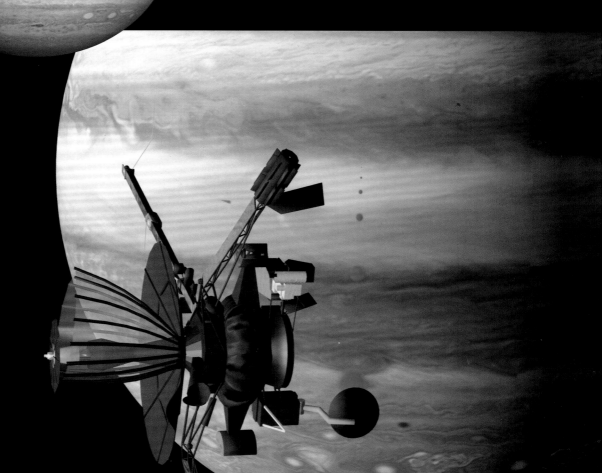

▲ Jupiter, with its Great Red Spot and showing detail in its belts and zones, photographed by the Hubble Space Telescope in April 2015.

▶ An artist's impression of how NASA's Galileo space probe looked as it orbited Jupiter from 1995 to 2003.

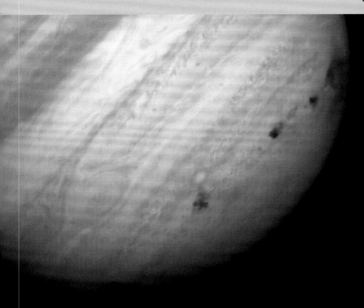

▲ The dark 'bruises' left on Jupiter's cloud tops by the impacting fragments of Comet Shoemaker–Levy 9 in July 1994 are imaged here by the Hubble Space Telescope.

GALILEO SPACECRAFT

With the Pioneer 10 and 11 and Voyager 1 and 2 flybys having provided tantalizing information about Jupiter, space scientists were keen to study its environment close-up over a longer timescale to learn more about the formation and evolution of the Solar System.

NASA's Galileo mission, named after the historic Italian astronomer, was launched from the cargo bay of Space Shuttle *Atlantis* in October 1989. Galileo, bundled with a smaller probe, took the long route to Jupiter, via Venus and then Earth twice, to gain energy that would propel it like a slingshot to its ultimate destination. Another bonus was a close pass of two asteroids, Gaspra and Ida.

Before Galileo even reached Jupiter, it became uniquely placed to watch as fragments of a comet pulled apart by the planet's gravity hit its cloud tops. The impacts by Comet Shoemaker–Levy 9 left obvious 'bruises' that could be seen with Hubble and other telescopes when they turned into view.

In July 1995, Galileo released its companion probe which, five months later in December, dived directly into Jupiter's atmosphere at 170,000 km/h. It then parachuted

153 km into the clouds, gathering weather data that it transmitted via Galileo back to Earth. The probe measured winds blowing at 725 km/h before it was destroyed by Jupiter's intense heat.

Galileo itself fired its engine to put itself into a highly elongated orbit around Jupiter to become the planet's first man-made satellite. This allowed it to spend time close to Jupiter, but also to fly close by a number of its moons. Mission extensions saw it eventually spend nearly eight years studying the planet and its family of moons.

Galileo discovered that there is much less water in Jupiter's atmosphere than had been expected, but a similar level of helium to that in the Sun. It also found a new and intense radiation belt about 50,000 km above the planet's cloud tops.

Close observations of Io showed that continuing volcanic activity was resurfacing the innermost moon. And perhaps most intriguingly, it collected evidence for an ocean of liquid salty water beneath the surface of Europa, and possibly Ganymede and Callisto too. To avoid any risk of contaminating Europa, Galileo was sent diving to destruction into Jupiter in 2003.

▲ An artist imagines the view from the surface of Europa, one of the Galilean moons of Jupiter. A plume of water vapour erupts from the ocean beneath its crust.

JUPITER – THE STATS

Diameter at equator: 142,984 km
Mass: 318 × Earth
Rotation period (day): 9 hours 50 minutes
Mean distance from Sun: 778,330,000 km
Orbital period (year): 11.86 years
Number of moons: 67

GALILEAN MOONS

Jupiter has a huge collection of moons – 67 at the last count – but most are small, leftover chunks from the formation of the Solar System that were caught by its gravitational embrace. Only four have significant size and they are the Galilean moons, so called after the Italian astronomer who spotted them through his new telescope in 1610. In order outwards from the planet, they are Io, Europa, Ganymede and Callisto. They were named by Simon Marius, a German astronomer, who claimed to have spotted them shortly before Galileo but failed to publish his observations. The names themselves were suggested by Johannes Kepler, famed for explaining in the 17th century the laws that determine how planets move in their orbits. Early observations of the timings of eclipses of Io by Jupiter's shadow allowed the first estimates of the speed of light by Ole Rømer at the Royal Observatory, Paris. But most of what we know about the moons themselves has come from visiting space probes.

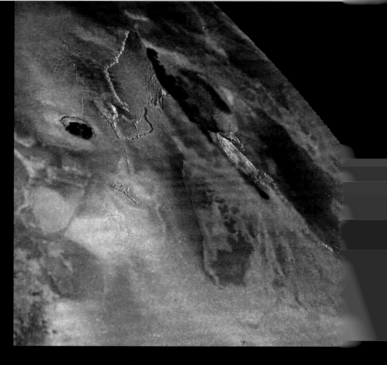

▲ A close-up of Io from the Galileo probe captured a terrain including gas vents, volcanic peaks, hot flowing lava (shown in red), and cooling lava (shown in black).

IO

Jupiter's innermost Galilean satellite came as a big surprise when the Voyager probes first photographed it close-up. It was revealed to be a highly active volcanic world, whose explosive nature made it resemble a pizza. Voyager 1 imaged a volcano, dubbed Pele, in the act of erupting when it flew by in March 1979. Four months later, when Voyager 2 sped by in July, Pele had stopped erupting but another was belching its sulphurous deposits to add to the 'pizza topping'.

When the Galileo spacecraft arrived, it imaged many new and colourful deposits showing how quickly Io's surface is replenished by plumes that were seen to shoot 300 km high. The lava's temperature was measured to be as high as 1,700°C, which is hotter than found in Earth's volcanoes, and suggests that the material comes from deep within Io. The volcanic activity is driven

by the powerful tidal pull that Jupiter exerts on Io, churning up its interior and causing its surface to bulge. Europa and Ganymede also have an influence here.

Io has a diameter of 3,630 km, making it slightly larger than the Moon, and orbits the planet once every 1.77 days, at a mean distance of 421,800 km, keeping the same face towards it. Around a ton of material from Io's surface is stripped away by Jupiter's magnetic field every second.

▼ The diverse nature of Europa's surface features includes blue and white areas of water ice, crisscrossed by cracks and ridges made up of other material

▲ Jupiter's four Galilean moons are shown here in order from the planet and to scale. From left to right, they are Io, Europa, Ganymede and Callisto.

EUROPA

The next main moon out from Jupiter could hardly be more different from Io. Europa is a rocky world that features no mountains and virtually no cratering. Instead its smooth, icy surface was shown by the Voyager probes to be covered with a surprising pattern of cracks like in a shattered eggshell. The pattern resembles ridge ice found in the Arctic Ocean, and is thought again to be caused by the tidal forces exerted by Jupiter and also other moons. Even more interestingly for the planetary scientists, the Galileo probe's observations of Europa indicate that it is entirely encircled by a salty sea just beneath its rocky crust. This subterranean ocean may be more than 100 km deep, which would mean that it holds more water than all Earth's seas combined. The water is kept liquid by the powerful pull of Jupiter, raising the intriguing question of whether this is an environment where simple life might have flourished. A future mission is being planned to find out!

Europa is the smallest of the Galilean moons, with a diameter of 3,140 km and orbiting at a distance of 671,100 km, once every 3.55 days. It keeps the same face towards Jupiter, being tidally locked, and Galileo has confirmed that there is a highly tenuous atmosphere, or exosphere, mainly of oxygen. At Europa's heart is an iron core.

▼ An artist imagines the view from Europa's icy surface, tinged with other areas of reddish material, as the giant planet Jupiter looms over the horizon, illuminated by a distant Sun.

THE REST OF THE FAMILY

▶ Callisto's dark surface is peppered with bright scars from a long history of impacts.

GANYMEDE

The third of the four Galilean satellites from Jupiter is the largest moon in the Solar System – Ganymede. With a diameter of 5,262 km, Ganymede is bigger even than planet Mercury. The Voyager probes revealed light and dark regions, including a huge dark area, named Galileo Regio. Subsequent imagery from the Galileo spacecraft shows that Ganymede has an icy surface, and that ridges and grooves divide a number of dark zones peppered with impact craters. Beneath this icy, rocky crust there is believed to be another ocean of water as well as an interior of rock and metal. Unusually, Ganymede also has its own magnetic field, thought to be generated by liquid iron around its core, which produces auroral displays. The Hubble Space Telescope discovered that Ganymede also is surrounded by a tenuous exosphere. The moon orbits Jupiter once every 7.16 days at a distance of 1.07 million km.

CALLISTO

Jupiter's outermost Galilean satellite, Callisto, is the third largest moon in the Solar System, after Ganymede and Saturn's Titan. Its distance has left it unperturbed by the tidal forces of Jupiter and the interplay between its three big neighbours. Its dark terrain therefore shows virtually no signs of renewal, and its surface is the oldest and most heavily cratered known, and essentially unchanged over the last 4 billion years. Features include two large ringed formations, first spotted by the Voyager probes, and named Asgard and Valhalla. Callisto, with a diameter of 4,800 km, orbits Jupiter once every 16.69 days at a distance of 1.88 million km, which is safely removed from the lethal zone of radiation from the giant planet. It might therefore make a suitable base for astronauts when humans eventually reach the outer Solar System. The Galileo probe showed that Callisto's interior is a mix of ice and rock, with a possible liquid ocean about 200 km underground. There is a thin exosphere.

▲ Ganymede's surface, imaged by the Galileo spacecraft, includes older, darker, cratered regions and lighter areas that have been shaped more recently. The bright spots are from material ejected by impacts in geologically recent times.

SATELLITE STRAGGLERS

Until spacecraft visited Jupiter, only 13 moons were known in total, but the number has since soared. All are very much smaller than the big Galilean four. Most, if not all, must be captured asteroids and 47 are less than 10 km across.

Eight are regular bodies in circular orbits, while the remainder are irregular satellites with tilted, elongated orbits. The most distant known, labelled S/2003 J 2, lies around 29.5 million km from the planet.

The largest moon of this mixed bunch is Amalthea, which is just 168 km wide and is one of four that lie within the orbit of Io. The others are Metis, Adrastea and Thebe. Amalthea was the fifth moon discovered, though not until 1892. Four other moons, Leda, Himalia, Lysithea and Elara, make up the Himalia group which follow similar orbits in the same direction that Jupiter rotates. Those orbiting the opposite way include the Carme, Ananke and Pasiphae groups.

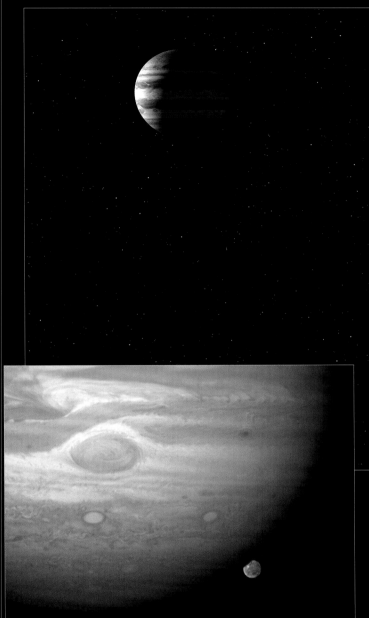

▲ Ganymede has its own magnetic field which produces auroral displays, pictured here by an artist, with giant planet Jupiter in the background. The aurora's motion supports claims for the existence of an underground ocean.

◄ Ganymede is caught slipping behind mighty Jupiter in this image taken by the Hubble Space Telescope. The Great Red Spot is prominent in the giant planet's cloud bands.

DELICATE RINGS OF DUST

One of Voyager 1's discoveries as it flew past Jupiter in February 1979 was that the giant planet is encircled by its own system of rings. Unlike the spectacular rings of outer neighbour Saturn, these are very delicate structures made up of fine dust ejected by meteoroid impacts on the inner moons. The rings are in three parts, a pair of faint outer bands called the Gossamer Rings, and a thick inner component called the Halo. Two tiny moons, Adrastea and Metis, orbit Jupiter within the inner Gossamer Ring, while the outer ring is fed dust by Amalthea and its neighbour Thebe.

▶ An illustration shows the layout of Jupiter's faint ring system

MISSIONS TO JUPITER

Space scientists use Jupiter's powerful pull to boost space probes on their journeys to the outer Solar System. Since Galileo, two other missions to more distant worlds have taken advantage of their Jupiter flybys to tell us more about the planet.

CASSINI–HUYGENS

Cassini–Huygens is a joint American–European mission that has been very successful in exploring Saturn. But at the tail end of 2000, it told us more about Jupiter as it sped by. Around 26,000 images were taken over a six-month period, giving detailed new views of Jupiter's cloudy features, its moons and rings. The spacecraft's studies also gave scientists a fresh understanding of how Jupiter's atmosphere circulates, revealing that gas was rising in the dark belts and falling in the bright zones, contrary to what had previously been thought. The irregular shape of particles in the rings showed that they had probably been produced by meteoroid impacts with the inner moons.

▲ The Galilean moon Io floats above Jupiter's cloud tops in an image taken by NASA's Cassini–Huygens probe on its way to Saturn in December 2000.

NEW HORIZONS

NASA's New Horizons probe was the first mission to Pluto and was sent racing at record speed to reach the distant world in less than ten years, arriving in 2015. *En route*, it flew past Jupiter at a distance of 2.3 million km, getting a speed boost. For four months around closest approach it measured activity in Jupiter's atmosphere, including a new storm dubbed the Little Red Spot. It also took fresh images of the rings and Galilean satellites, including volcanoes erupting on Io.

JUNO

Arriving at the giant planet in July 2016 was NASA's latest probe, Juno, named after the Roman goddess who was Jupiter's wife and queen. The mission, launched in August 2011, was planned to focus on Jupiter itself, flying in close polar orbits to examine the planet in detail in a bid to learn how it formed and evolved. Jupiter is

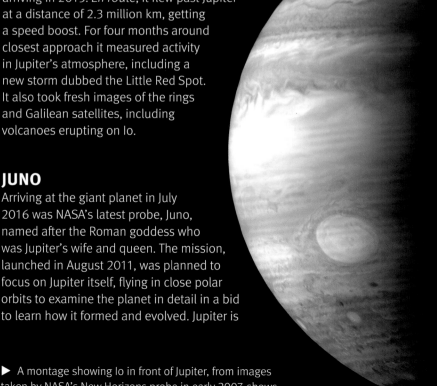

▶ A montage showing Io in front of Jupiter, from images taken by NASA's New Horizons probe in early 2007 shows

made up of most of the material that was left in the Solar System after the Sun formed, so will help provide the recipe for planets.

Juno took the now familiar scenic route via the inner Solar System to gain momentum, flying back past Earth in October 2013. But on reaching Jupiter it was sent into an orbit quite different to Galileo's, circling from pole to pole and slowly spinning to allow an array of instruments to scan the planet, including a powerful colour camera. Its orbital pattern means that over 37 planned orbits, each lasting 11 days, Juno will be able to map the whole planet's atmosphere and penetrate deep into it with its experiments.

The low elliptical orbit was designed to bring Juno to a height of just 5,000 km above the cloud tops. Hugging the planet in such a way helps it to avoid the highest levels of radiation pumped out by Jupiter. However, over a year they are still equivalent to 100 million dental X-rays, so Juno's sensitive electronics were shielded within a 1 m wide titanium vault with walls a centimetre thick.

Juno's mission is due to end in February 2018 when it will, like Galileo, be sent diving to destruction into the planet's atmosphere.

JUICE

Europe is busy working on a new mission to visit Jupiter's satellites. JUICE – short for JUpiter ICy moons Explorer – is due to launch in 2022 and reach the giant planet in 2030. As well as studying Jupiter itself, JUICE will examine the moons Europa, Callisto and Ganymede, and the planet's rings. It will

▲ An artist's impression of ESA's JUICE spacecraft, which is set to begin exploring Jupiter and its satellites from 2030 onwards.

then become the first spacecraft ever to go into orbit around Jupiter's largest moon, Ganymede. Scientists want to know more about this moon's interior and how it comes to have a magnetic field of its own. The mission also aims to find out whether Europa and Ganymede have the water, heat, organics and stable conditions to allow life to have flourished.

MISSION TO EUROPA

NASA is currently planning its own mission to the moon Europa, which is seen as possibly the most likely environment in which to find extraterrestrial life in the Solar System. A probe shielded against Jupiter's dangerous radiation will fly in a long orbit that brings it swooping low over Europa several times, as low as 25 km above its surface and imaging it in high resolution. Other instruments will investigate the structure and composition of Europa's interior and icy shell, using a thermal camera and ice-penetrating radar. The mission is scheduled to launch in the next decade.

▲ How NASA's latest space probe Juno might have looked as it arrived in orbit around Jupiter in July 2016 to begin orbiting from pole to pole, making the closest study yet of the giant planet to

SATURN – THE RINGED BEAUTY

Saturn, the second largest of the planets in our Solar System, is unlike any other object we can see. That is because it is encircled by a system of bright rings that give it a unique appearance. Since the start of the Space Age, astronomers have detected rings around other worlds too, in particular Jupiter and Uranus, but they are faint and bear no comparison to Saturn's splendid and beautiful array. NASA's orbiting Cassini spacecraft has been putting Saturn, its rings and family of 62 satellites under close scrutiny since 2004. It also landed a probe on largest moon Titan.

Despite Saturn's vast size, it is a considerably less dense gas ball than its closest rival, Jupiter. Though it could contain more than 760 Earths, it is only 95 times more massive than our own world and an eighth its density. If you could place the planet in a large enough sea of water, it would float.

Saturn's straw-coloured cloud tops show much less variation than the bands and belts on Jupiter, with their swirling weather patterns. But it does occasionally break out in spots, and one prominent white feature appears to recur in its northern hemisphere every 30 years or so, during that region's summer. As on Earth, the seasons are caused by the tilt of Saturn's axis, which amounts to 26.7 degrees.

Winds in the upper atmosphere blow much faster than those observed on Jupiter, reaching speeds of up to 1,800 km/h, nearly five times faster than any measured on Earth. A zone in Saturn's southern hemisphere was dubbed Storm Alley by

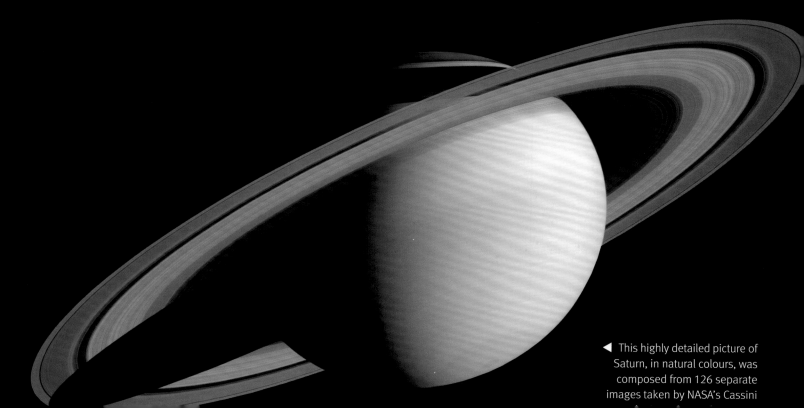

◀ This highly detailed picture of Saturn, in natural colours, was composed from 126 separate images taken by NASA's Cassini

▲ An unusual view of Saturn as it blots out the Sun and the rings scatter the sunlight. Just above and to the left of the bright main rings can be seen a tiny dot that is Earth.

NASA scientists. Observations from Earth of one hurricane there in 2007, thousands of kilometres wide, showed it to be still raging months after its discovery, and it was followed by an even bigger tempest in 2010 that encircled the planet and lasted a year.

Huge storms have also been seen to be raging over Saturn's north and south poles. The one at the north pole is more than 8,000 km wide, around two-thirds the diameter of the Earth, with winds blowing clockwise at 560 km/h. The southern polar storm is an even bigger tempest, being about 30,000 km across and 100 km deep. First spotted by NASA's Voyager probes, it has a curious six-sided shape like a hexagon, which may be due to shallow jets of wind disrupting the main wind flow.

Similarly to Jupiter, Saturn is mainly made up of hydrogen and helium, in the proportions of about three-quarters hydrogen and a quarter helium. There are also small amounts

SATURN FACT FILE

- Saturn is named after the Roman god of agriculture.
- Like Jupiter, Saturn has a squashed appearance due to its rapid rotation. Its polar diameter is 11,810 km less than at the equator.
- Aurorae seen in Saturn's atmosphere are thought to be caused by charged particles from the solar wind meeting its magnetic field, which is 578 times stronger than our own.
- Lightning bolts 10,000 times more powerful than on Earth have been recorded in its storms.

of water, methane and ammonia. Though it lacks a solid surface, there is believed to be a small solid core of rock and ice at its very heart, wrapped in a layer of liquid hydrogen. Temperatures are a cool −175°C at the cloud tops, but rise rapidly as one penetrates the interior, reaching 11,700°C at the centre. This intense heat is thought to be caused by helium sinking through the hydrogen. It means that Saturn radiates about 2.5 times as much energy as it receives from the Sun.

Churning material deep within Saturn create a 'protective bubble' magnetosphere around the planet that is much bigger than Earth's. And unlike ours, the magnetic poles are closely aligned with the poles of its rotational axis.

SATURN – THE STATS

Diameter at equator: 120,536 km
Mass: 95 × Earth
Rotation period (day): 10 hours 40 minutes
Mean distance from Sun: 1.429 billion km
Orbital period (year): 29.46 years
Number of moons: 62

SATURN'S SPECTACULAR RINGS

Seven main rings encircle Saturn, and the entire system measures 282,000 km from one extreme to the other, though they are typically only 10 m thick. Close-up images from space probes show that there are actually thousands of separate ringlets, formed of countless particles of rock and ice from shattered moons, comets and asteroids that came too close.

Saturn's unique appearance was first noticed by Galileo Galilei in 1610, from Italy, using one of the earliest telescopes, though he failed to understand why it looked different. The instrument was not good enough to allow him to recognize the rings, and he thought at first that the planet was formed of three separate spheres before deciding the rings were a pair of handles. Their ringlike nature was observed by Dutch astronomer Christiaan Huygens, who studied Saturn from 1655 with a more powerful telescope. He also sketched the rings' shadow on the planet. Huygens thought that the rings were a solid disk. Showing great foresight, a French astronomer, Jean Chapelain, suggested in 1660 that they were actually a swarm of small satellites. We know now that they are made from billions of particles of all sizes.

An Italian astronomer called Giovanni Cassini was the first to notice a gap within the rings, and this became named after him as the Cassini Division. It is about 4,800 km wide. Another division, about 325 km wide, was spotted by James Keeler in 1888 and is known as the Encke Gap. Many smaller ones have since been found. They are caused by the shepherding of particles by various smaller satellites of Saturn, just as sheepdogs control the movement of flocks

▼ A detailed but simulated picture of Saturn's ring structure, constructed by measuring radio signals sent through them by Cassini from the other side.

▲ A view down on to Saturn from high above shows the planet's shadow against the system of rings.

of sheep. One of Saturn's larger moons, Mimas, controls the Cassini Division, and a smaller one, Pan, orbits within the Encke Gap. The rings, which are labelled after letters of the alphabet, orbit Saturn at different speeds, and they contain structure, like a clumping together of icy snowballs, which is due to the influence of satellites but is not yet fully understood.

NASA's Cassini spacecraft has studied Saturn in unprecedented detail from orbit around the planet, but to enter its orbit it had to fly through the gap between the F and G rings. It was an anxious time for space scientists, and the probe's cameras and instruments were switched off to keep them safe during the passage. However, no harm was done and the spacecraft survived.

RINGING THE CHANGES

Saturn's rings do not always look the same. Because the planet is tilted against the plane of its orbit, they appear to us to open and close as it travels around the Sun. When one or other pole is leaning towards the Sun, during local summer or winter, the rings appear open as wide as we can see them, but at Saturn's equinoxes we get a sideways view of the planet and see the rings edge-on. Galileo was first to spot this phenomenon, getting a shock when he pointed his telescope at Saturn in 1612 only to find that its 'handles' had disappeared.

If we could view Saturn's rings really close up, we would see that they are formed of clumps of particles of rock and ice, similar

OBSERVING SATURN

To the unaided eye, Saturn resembles a bright star, though it does not twinkle because it is not a starlike point of light. The smallest telescope, even with low magnification, is enough to show it as a disk and to reveal those rings, marking Saturn out as a celestial jewel and one of the finest observing targets for an amateur astronomer. Even binoculars will show there is something odd about the shape. A small telescope, with an aperture of at least 75–100 mm, and steady skies, will also be enough to spot the Cassini Division in the rings, and the brightest of Saturn's moons.

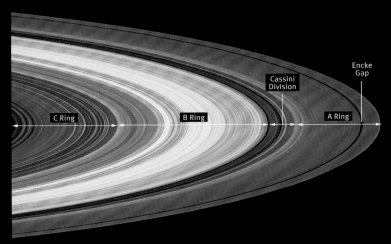

▲ A picture in false colour of Saturn's main rings, made after Cassini monitored fluctuations in starlight shining through them.

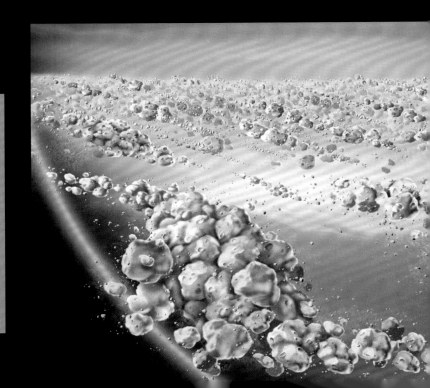

THE CASSINI MISSION

Since 2004, a robotic probe called Cassini has been orbiting Saturn, bringing space scientists a wealth of information about the planet, its rings and family of moons. The mission was a collaboration between the United States and Europe, with NASA providing Cassini and the European Space Agency supplying a piggyback probe called Huygens to investigate its largest moon, Titan. Both spacecraft were named after historic observers of the planet.

Cassini–Huygens was launched in 1997 and took the now common indirect route to get to Saturn, flying past Venus, Earth and Jupiter to boost it on its way. No sooner had it arrived at the Saturn system in late 2004 than it released ESA's Huygens, which safely soft-landed on the surface of Saturn's largest moon Titan on 14 January 2005, sending back incredible pictures.

Meanwhile, Cassini was beginning to orbit Saturn, collecting data with its range of instruments. Some would image the planet and its moons in different types of light, some invisible. Others studied the dust, charged particles and magnetic field around the planet. Radar and other sensing using microwaves allowed the spacecraft to investigate the atmosphere, analyze particles in the rings, check the mass of moons and map the surface of Titan.

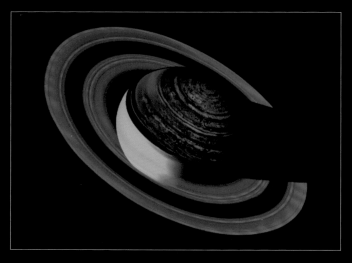

▲ This colourful picture was produced by combining Cassini's views of Saturn in different wavelengths. The icy rings appear blue while red represents warmth radiating from the night side.

Cassini's mission was originally designed to last for four years but has been extended a number of times since and the spacecraft continues to function well, sending home much useful information about the Saturn system. As the mission nears its end, the science team have been taking greater risks such as getting it to fly lower into jets of spray from an active moon, Enceladus.

▶ An artist's depiction of how NASA's Cassini probe looks in orbit around the giant, straw-coloured world Saturn.

Eventually, the truck-sized probe will be steered into Saturn's atmosphere to burn up, but before then it will fly between the planet and its innermost ring 22 times to gain new data.

During its extended lifetime, Cassini has made some extraordinary discoveries. The first was the close-up views of Titan as Huygens made the most distant landing ever achieved in the Solar System. Subsequent studies showed Titan to be an Earth-like world with lakes, rivers and rain.

Then plumes of icy material were observed by the probe to be bursting from cracks around the south pole of moon Enceladus. When water ice was found in the spray, it raised the intriguing question as to whether this moon might harbour life in a subsurface ocean. Cassini swooped low into the plumes to get a taste of what they contained.

Cassini was able to study close-up for the first time the dynamic processes that were happening within Saturn's rings, including propeller-like formations, vertical structures, and the possible formation of a new moon.

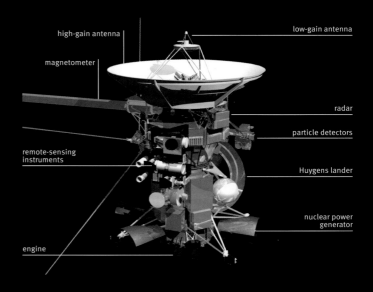

▲ Scientific instruments and other components of the Cassini–Huygens probe are indicated in this detailed diagram.

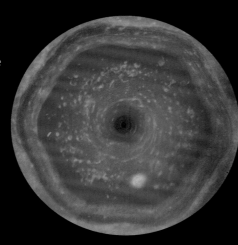

▶ Cassini scientists assembled several images to produce the highest-resolution view of the peculiar six-sided feature at Saturn's north pole resembling a hexagon.

A huge storm erupted in Saturn's northern hemisphere late in 2010, violently distorting the usual calm nature of the cloud bands. It grew to encircle the planet within months, and only subsided after its head collided with its tail a year later. Cassini detected molecules in the storm that had not previously been seen in the upper atmosphere and the biggest temperature increases ever recorded on any planet.

Cassini also has given space scientists close-up looks at the hurricanes raging at the north and south poles, including a detailed view of the northern storm's peculiar hexagonal pattern.

▼ Cassini–Huygens blasts off from Cape Canaveral, Florida, on a night-time launch to begin its seven-year journey to the planet Saturn, via Venus, Earth and Jupiter.

CASSINI FACT FILE

- Cassini was launched by a Titan IVB/Centaur rocket from Cape Canaveral, Florida, on 15 October 1997.
- The orbiter is 6.7 m high and 4 m wide, about the size of a school bus. Its Huygens lander was 2.7 m in diameter.
- Cassini travelled 3.5 billion km in a circuitous route across the Solar System to reach Saturn. It is powered by a small nuclear generator.
- Saturn's average distance from Earth is 1.43 billion km and Cassini's signals take between 67 and 85 minutes to reach Earth.

TOUCHDOWN ON TITAN

One of the triumphs of the Cassini mission came early on when ESA's piggyback probe Huygens made a successful landing on the surface of Saturn's largest satellite, Titan. It was the first landing on a planetary moon other than our own and sent back a picture of a rock-strewn landscape.

Titan, the second largest moon in the Solar System, was discovered as a starlike point close to Saturn by Dutch astronomer Christiaan Huygens in March 1655. It can easily be seen in binoculars. Only Jupiter has a bigger moon, Ganymede, and with a diameter of 5,150 km, Titan is even larger than innermost planet Mercury, and is the only moon to have a substantial atmosphere. It is even denser than the Earth's atmosphere.

Today we know that Titan is tidally locked, just like our own Moon, so that it keeps the same face towards Saturn throughout every orbit, which lasts 15 days 22 hours.

Before Cassini, the best views of Titan were gained by NASA's Voyager space probes in 1980 and 1981. But they showed just an orange, hazy ball as the probes were unable to see through the thick petrochemical smog. Better results were obtained by the Hubble Space Telescope, whose infrared camera picked up hints of two surface features that were dubbed Xanadu and Shangri-la.

▲ Saturn's biggest moon, Titan, is pictured in front of the planet and its rings, seen edge-on. The rings' shadows are visible on the cloud tops.

Having been carried by Cassini throughout the long voyage, Huygens was detached on 25 December 2004, as the mothership went into orbit around Saturn. On 14 January 2005, Huygens plunged into Titan's nitrogen-rich atmosphere and began its descent, lasting 2 hours 27 minutes, by parachute to the surface, recording the sound of the wind as it went.

The wealth of data collected by Huygens' instruments revealed that the moon had many Earth-like features, including rivers, deltas and shorelines. When the probe finally hit the ground, its crunch told scientists that they were on a spongy

▼ A mosaic of near-infrared images taken by NASA's Cassini spacecraft pierced Titan's hazy atmosphere to catch sunlight glinting off its seas and lakes.

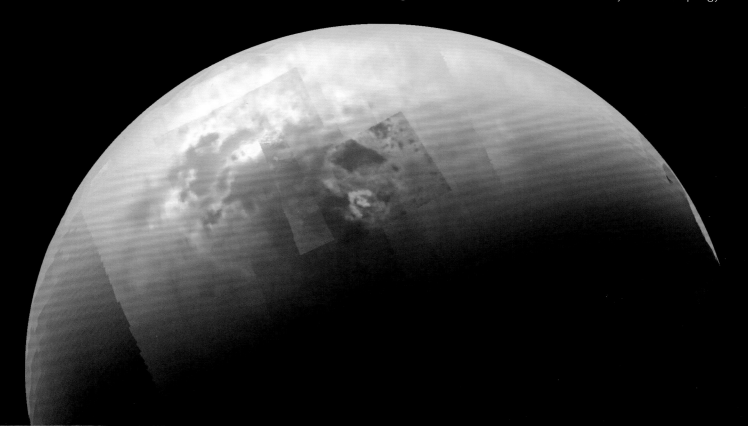

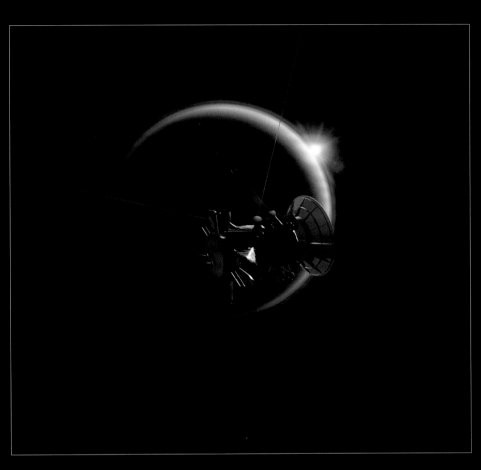

▲ Rocks appear rounded like the pebbles on a beach in this first colour view from the surface of Titan, taken by Huygens.

▲ An artist imagines the Cassini probe passing over the night side of Titan, with the distant Sun visible against the moon's bright limb.

surface like wet sand, which appeared to be a dry lake bed. A photo of the surroundings showed rocks, but they were rounded like pebbles on a beach, suggesting they had been eroded by liquid. But it is not water that is responsible but ethane and methane, kept liquid by a surface temperature of −180°C.

Winds blow on Titan just like on Earth, and clouds form, dumping showers of walnut-sized petrochemical raindrops. Cassini–Huygens found that the moon's atmosphere is full of complex hydrocarbons caused by methane being broken apart by the sunlight. Scientists believe that frozen Titan resembles Earth as it was billions of years ago before life formed and began to fill the atmosphere with oxygen.

Huygens did not survive for long, but Cassini carried on studying Titan during more than 70 flybys, mapping much of its surface. It observed hydrocarbon sand dunes, mountain ranges, many lakes of methane, and possible ice volcanoes. The surface features appear relatively young, suggesting they have been reshaped by geological processes within the last billion years.

The lakes of Titan have been known to fill and dry out as the seasons change, and huge clouds were seen forming in the southern hemisphere in recent years as the region moves into winter. Titan has three large seas, called Kraken Mare, Ligeia Mare and Punga Mare, plus numerous lakes. The changing shape of some coastal features has intrigued scientists who are keen to send another mission to Titan, possibly including a robotic yacht or submarine.

▶ Huygens' entry into the dense atmosphere of Titan and subsequent descent by parachute is illustrated by an artist.

SATURN'S MANY MOONS

Saturn has a vast family of moons, with 62 at the last count. But all since the tenth discovery – Janus, in 1966 – have been found during the Space Age. Including Titan, seven are massive enough to have become spherical in form – in order outwards from the planet, they are Mimas, Enceladus, Tethys, Dione, Rhea, Titan and Iapetus.

Following Dutch observer Christiaan Huygens' discovery of Titan in 1655, Italian astronomer Giovanni Cassini discovered four of the moons – Iapetus, Rhea, Dione and Tethys – using an early telescope, between 1671 and 1684. A century later, German-born astronomer William Herschel spotted Mimas and Enceladus in 1789, from England, with a home-made telescope. All seven satellites were named by his son, John Herschel.

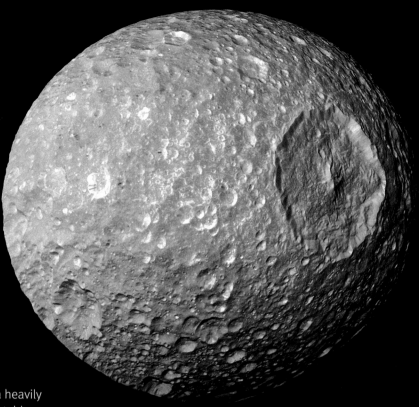

MIMAS

The closest and smallest of the big seven, Mimas is a heavily impact-scarred world just 418 km in diameter. It is notable for having one giant crater that is almost a third the satellite's diameter. Named after the moon's discoverer, Herschel crater makes Mimas look remarkably like the Death Star in the *Star Wars* movies. The impact that produced it must almost have broken this moon apart.

▲ Mimas seen up close reveals its many scars including the giant one to the upper right, named Herschel crater, which was caused by a huge impact.

ENCELADUS

Enceladus is an icy moon, 512 km across, that has intrigued scientists thanks to the salty spray that can be seen erupting from deep cracks, dubbed 'tiger stripes', around its south pole. The source appears to be an underground sea, and the Cassini probe's studies suggest the ocean surrounds the entire moon, with the crust completely detached from the solid interior. These fissures are caused by the tidal stresses that Enceladus experiences when orbiting Saturn, and the water ice jets out like a shaken-up fizzy drink. The ocean, which is about 40 km deep, is kept liquid by gravitational churning.

Organic material that forms the building blocks of life has been detected already in the geysers, leading space scientists to wonder whether the ocean might be one of the best places to look for alien life. At the tail end of 2015, Cassini risked making three very close encounters to get a taste of the jets from Enceladus.

TETHYS

Tethys, which is 1,066 km in diameter, also has a large but shallow crater, 400 km wide, called Odysseus. But it is less heavily scarred than some of the other moons, possibly because the tidal warming from Saturn kept it from solidifying for longer. It also has an ancient grand canyon called Ithaca Chasma, which stretches for 2,000 km and is 100 km wide and up to 5 km deep.

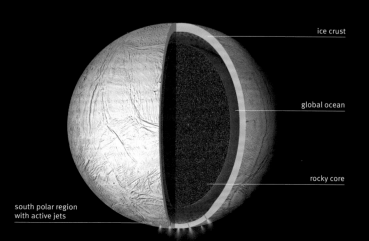

ice crust

global ocean

rocky core

south polar region with active jets

◀ A cutaway diagram of icy Enceladus reveals how scientists believe the moon to be encircled completely by an ocean between the crust and the solid centre. (The layers are not drawn to scale.)

TETHYS

DIONE

IAPETUS

DIONE

The Voyager 1 spacecraft recorded wispy markings on Dione's surface, which led scientists to wonder whether they were flows of volcanic ice. Close studies by the Cassini probe confirm them to be cliffs of ice, some several hundred kilometres high. The pattern of cratering on Dione leads planetary experts to believe that it was spun round by a cosmic impact some time in the past.

RHEA

Saturn's second largest moon, with a diameter of 1,528 km, is one of the most heavily scarred worlds in the Solar System, and includes two ancient impact basins. Again there are icy cliffs, but the moon could have been geologically active in the distant past. Bright rays from young impact craters lead scientists to think that the present surface is around 4 billion years old. Rhea also has its own faint rings surrounding it.

IAPETUS

The Cassini mission finally solved a mystery about this 1,450 km diameter moon that had existed since its discovery. It appeared much darker when on one side of Saturn than on the other. Today we know that Iapetus really does have a dark side, due to a smaller moon, Phoebe, dumping material on it. It also has a strange mountain ridge running around the equator.

BEST OF THE REST

Perhaps the most eye-catching moon in Saturn's family is Hyperion, discovered in 1848. It is only 355 km across but its deep craters make it look like a bath sponge!

HYPERION

◄ Subtle colour differences of features on the moon Rhea have been enhanced in this close view from Cassini. The moon's surface is actually fairly uniform.

URANUS – COLD AS ICE

Uranus was the first planet to be discovered, because the five inner worlds had been known throughout history, due to their brightness. The discovery was made in 1781 by the famous German-born astronomer William Herschel, who had made England his home. He thought he had found a comet, but follow-up observations revealed it to be twice as far away as Saturn, taking 84 years to orbit the Sun. Herschel wanted to call his world Georgium Sidus to please the British king, George III, but sense prevailed and it was given a classical name, after the Greek god of the sky.

Uranus' disk proved frustratingly bland and devoid of detail. But early observers spotted its brightest moons. Herschel himself found two, Oberon and Titania, in 1787, and William Lassell found two more, Ariel and Umbriel, in 1851. Before the Space Age, only one more was found, Miranda in 1948 by US astronomer Gerard Kuiper. Today 27 are known, thanks to NASA's Voyager 2, the Hubble Space Telescope and powerful new instruments on Earth. Most are named after characters in Shakespeare's works.

The steep orbits of the moons revealed there was something very odd about Uranus. Its axis is tilted by nearly 98 degrees to the plane of its orbit, which means it is effectively rolling around the Sun on its side. Like Venus, it also rotates in the other direction to the rest of the planets. A huge impact with another body billions of years ago is thought to have knocked Uranus for six.

A big surprise for astronomers in 1977 was the discovery that Uranus has rings, just like Saturn, but much fainter. They revealed themselves by causing a star to blink several times as Uranus was passing in front of it, an event called an occultation. Voyager 2 and the Hubble Space Telescope later brought the number of known rings to 13.

▲ When Voyager 2 flew past Uranus in January 1986, it imaged a world that showed almost no features in its icy clouds.

The planet's unique orientation means it experiences extreme seasons. Each lasts around 21 years. When one hemisphere is experiencing midwinter, it is almost completely pointed away from the Sun and receives no sunlight. Winter is followed by a similarly long spring, when both hemispheres enjoy sunshine and night during its 17-hour-long day, before the other hemisphere is plunged into many years of darkness.

Uranus' northern hemisphere was in the deep grip of winter when the Voyager 2 spacecraft flew past in 1986, the only probe to visit the seventh planet. The spacecraft could see little going on within its clouds. But years later, with the onset of spring, and with some latitudes receiving sunlight for the first time in many years, huge storms were triggered that could be observed with powerful instruments on Earth. Following the spring equinox in 2007, the planet's southern hemisphere is now approaching midwinter.

Uranus puzzles space scientists because it is the coldest of the planets, with temperatures around −224°C in the cloud tops. It emits very little energy of its own, but the extreme seasons have a strong effect on its atmosphere. Some of the fastest winds in the Solar System have been detected, blowing at 800 km/h.

URANUS — THE STATS

Diameter at equator: 50,724 km
Mass: 14.5 × Earth
Rotation period (day): 17 hours 14 minutes
Mean distance from Sun: 2.877 billion km
Orbital period (year): 84.01 years
Number of moons: 27

OBERON

TITANIA

Voyager 2 was able to record features on just a handful of Uranus' satellites, though it also discovered several smaller ones on its flyby.

Uranus and its outer neighbour Neptune are known as ice giants because they are generally composed of far more ice and rock than the gas giants Jupiter and Saturn. Uranus' atmosphere is mainly hydrogen and helium, with a small amount of methane that gives the planet its blue-green hue. There are also traces of water and ammonia. Uranus has a dense liquid core of water, methane and ammonia.

Like many planets, Uranus has a magnetic field, but its axis is tilted almost 60 degrees to the planet's rotational axis.

ARIEL

UMBRIEL

OBSERVING URANUS

Though it was unknown to the ancients, Uranus is actually just visible to the naked eye under clear dark skies. Binoculars will allow you to see it easily, though it will look like a star. A telescope is required to reveal its tiny blue-green disk.

MIRANDA

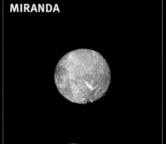

PUCK

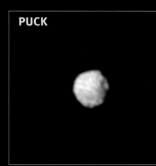

▲ Uranus' many faint rings are revealed in this photo taken by Voyager 2, as well as a few of its satellites among the stars.

URANUS FACT FILE

- Uranus is the only planet named after a Greek god. All the others were given names of Roman deities.
- It is the second least dense planet (after Saturn). Despite its size, its gravitational pull is less than that on Earth.
- Though most of Uranus' moons are named after Shakespeare characters, three – Ariel, Umbriel and Belinda – are from Alexander Pope's poem 'The Rape of the Lock'.
- The rings are formed of extremely dark particles. William Herschel reported observing a ring in 1797.

NEPTUNE – PLANET OF THE DEEP

Neptune was the second of the new planets to be identified, but it was not discovered by lucky observation, as Uranus had been. Instead, the eighth planet was located mathematically. Uranus was not behaving quite as expected in its orbit, and it was suggested that the pull of another planet was disrupting its path. A French mathematician, Urbain Le Verrier, weighed up this proposed world by calculating its mass and predicted the very area in the sky where it should be found. He managed to get a German astronomer, Johann Galle, interested in the project and he found it the first night he looked with a telescope at Berlin Observatory, on 23 September 1846. It was named after the Roman god of the sea.

Le Verrier's success was a blow to an English mathematician, John Couch Adams, who had reached similar conclusions and asked the Astronomer Royal, Sir George Airy, to search for the planet. He had been ignored. He finally got astronomer professor James Challis to begin a search at Cambridge Observatory, England, and the professor actually twice logged Neptune in August 1846, but failed to notice it. (Subsequently, it was found that Galileo had sketched Neptune close to Jupiter in 1613 but thought it was a star.)

Neptune is an ice giant, like Uranus, but has a stronger blue tinge due to having more methane in the atmosphere, along with its hydrogen, helium, ammonia and water. It also shows more features in its cloud tops. There is thought to be an Earth-sized solid core. Neptune is tilted by 28 degrees to its orbital plane, so that it experiences seasons, though far less extreme than its inner neighbour. Its magnetic axis is tilted 47 degrees to its polar axis, and has a magnetic field 27 times stronger than Earth's.

When Voyager 2 flew by, Neptune was displaying deep blue bands and a giant dark spot that marked a raging storm which has since vanished. The spacecraft also recorded bright clouds racing around the planet at 1,100 km/h.

◄▼ Images of Neptune, left and below, captured by the Voyager 2 spacecraft as it shot past the planet in 1989. The dark spot can be seen in both pictures.

NEPTUNE – THE STATS

Diameter at equator: 49,244 km
Mass: 17.1 × Earth
Rotation period (day): 16 hours 7 minutes
Mean distance from Sun: 4.498 billion km
Orbital period (year): 165 years
Number of moons: 14

▲ Neptune's largest moon Triton reveals its mottled surface – the greenish band is thought to be mainly fresh nitrogen frost.

NEPTUNE FACT FILE

- Neptune is the smallest of the four giant planets in the Solar System. However, it is more massive than Uranus because it is denser.
- The gravitational pull of Neptune is only a little greater than that on Earth, being about 17% stronger.
- Deep beneath the clouds that surround Neptune may be an ocean of super-hot water. The pressure at this level stops the water from evaporating away.
- Neptune experiences incredibly windy weather with gales blowing at more than 2,000 km/h.

moons, and it is believed to be an object captured from the Kuiper Belt beyond Pluto. It appears to be a rocky moon but is remarkable because, despite being one of the coldest objects in the Solar System, with a surface temperature of −235°C, it is volcanically active. Voyager 2 saw several spouting geysers and there are few craters on its crust of frozen nitrogen, suggesting it has been renewed. The surface resembles that of a cantaloupe melon. Triton has a thin atmosphere of nitrogen with a little methane, which is probably fed by the geysers.

Neptune's second discovered moon, Nereid, is only 340 km across but has the most stretched orbit of any known satellite in the Solar System. It could be a captured asteroid.

OBSERVING NEPTUNE

Neptune is too faint to be seen without optical aid, but binoculars will show it as a 'star' of about magnitude 8. Its slow crawl around the sky means it spends an average of 14 years in each constellation of the zodiac.

WATER-GOD WORLDS

Only 17 days after the discovery of Neptune, English amateur astronomer and brewery boss William Lassell found a moon, Triton. It was the only known satellite for 103 years, but since then another 13 have been discovered, including six by Voyager 2. They are named after deities linked to water. Voyager is the only spacecraft to have visited Neptune, flying past it in 1989. During the encounter, it confirmed that the planet also has faint rings, as suspected from ground-based observations. Six are known and they are like arcs rather than complete rings.

Most of Neptune's moons are small and insignificant, but Triton is a sizeable 2,705 km in diameter. It is big enough to have hydrostatic equilibrium, or, in other words, a round

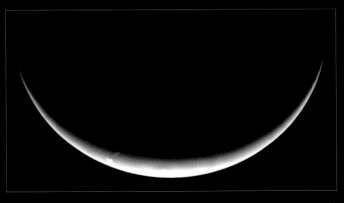

▲ Voyager 2 looked back after its encounter with Neptune to record the planet's night side, including the south pole. It then continued

PLUTO – OUT IN THE COLD

On 19 January 2006, a spacecraft launched from Cape Canaveral, Florida, on a mission to visit Pluto, the last of the classic planets in our Solar System. Ironically, only seven months later, with New Horizons' journey hardly begun, Pluto was demoted to a new status of dwarf planet by a gathering of astronomy's arbiters, the International Astronomical Union. When the NASA spacecraft eventually reached and shot past Pluto, ten years later, it found that regardless of how it is labelled, Pluto is one of the most astonishing worlds in the Solar System.

DISCOVERY OF PLUTO

Following the discovery of Neptune in 1846, astronomers wondered whether more planets were waiting to be found. One was Percival Lowell, the same man who had become convinced that there were canals on Mars. He began his own search from his observatory at Flagstaff, Arizona, but found nothing, though he managed to photograph Pluto twice without realizing it a year before his death in 1916.

Years later, a self-taught astronomer, Clyde Tombaugh, from Kansas, joined the observatory staff and took up the hunt. It was arduous work, which involved taking regular photos of the sky, processing them and then checking the thousands of stars for one that had moved. But less than a year after he began his search, Tombaugh found a candidate, and it was confirmed on follow-up photos. The discovery of Pluto was anno

◄ Pluto reveals it has a heart in this enhanced colour image from New Horizons.

The motive for searching for the new planet had been a remaining gravitational influence on Uranus that it was felt could not be accounted for by Neptune's pull. However, once Pluto was found it was realized that the discrepancy had been due to a misjudgement of Neptune's mass. Pluto was far too small to have had any influence in any case.

STRANGE NEW WORLD

Pluto's reclassification in 2006 was controversial, and was sparked by the discovery in the early years of the 21st century of other objects in its region of the Solar System, known as the Kuiper Belt. But even when Pluto was discovered, astronomers were unsure about its true nature. The telegram announcing its discovery termed it a 'trans-Neptunian body', and some

PLUTO – THE STATS

Diameter at equator: 2,374 km
Mass: 0.002 × Earth
Rotation period (day): 6 days 9 hours 17 minutes
Mean distance from Sun: 5,906 billion km
Orbital period (year): 248 years
Number of moons: 5

Signs that Pluto was different from the other planets included the nature of its orbit, which was tilted by 17 degrees to the plane of the ecliptic, which the other planets stuck close to. The orbit was also extremely eccentric, ranging from 7.4 billion km from the Sun when Pluto was at its most distant, to 4.4 billion km when closest, which brought it inside the orbit of Neptune.

When first found, Pluto was assumed to be at least as massive as Earth, but by 1976, when telescopic studies detected a surface coating of highly reflective methane ice, it was realized that it must be much smaller. The discovery of its largest moon, Charon, in 1978, allowed laws of physics to be applied to derive a diameter of 2,300 km, which is not far from the figure confirmed by New Horizons.

PLUTO'S MOONS

Discovered by observers from the US Naval Observatory, Charon was named after the ferryman in Greek mythology, who carried the dead across the river Styx in the underworld. Two more, much smaller, satellites were discovered by the Hubble Space Telescope in 2005, and named Hydra and Nix. Then, as New Horizons was racing to reach Pluto, another two tiny moons were found: Kerberos in 2011, and Styx in 2012. All of Pluto's moons are thought to have been formed when the dwarf planet collided with another body, billions of years ago.

◄ Sunlight shines through Pluto's hazy atmosphere, revealing its sky is blue, in a view taken as New Horizons looked back after its close encounter.

▼ An artist imagines how New Horizons looked as it shot past Pluto and largest moon Charon in July 2015, with the Sun resembling a distant, brilliant star.

PLUTO REVEALED AT LAST

New Horizons became the fastest spacecraft ever sent to the outer Solar System. In less than three months, it had already crossed the orbit of Mars, and just 13 months after launch it reached Jupiter. A close flyby, at a speed of about 23 km/sec, put the probe on a direct course for Pluto. Even at such speeds, it took New Horizons nine and a half years to make the whole 5 billion km journey. The spacecraft flashed by Pluto with pinpoint precision on 14 July 2015, taking spectacular images of the dwarf planet and its moon Charon, and gathering a wealth of data before it was quickly receding from Pluto again and heading ever deeper into space. The next few months were spent gradually sending a treasure trove of information home, with every signal taking four and a half hours to reach us. At last scientists had got a proper look at Pluto and they were astonished by what they saw.

▲ A close-up view of Pluto's largest moon Charon from New Horizons shows its red patch and the canyons running around the equator.

It had been imagined that Pluto might be a dead world, its ancient cratered surface covered in methane ice. But the truth was quite different. Pluto showed a variety of complex landscapes, and was far from uniform. Even as New Horizons began to get close enough to see detail, a bright plain came into view that was shaped like a heart. The left side, named Sputnik Planum, showed no impact cratering at all, suggesting it formed relatively recently. It is filled with nitrogen and carbon monoxide ices, constantly churning due to convection.

▼ Scientists pack the New Horizons mission control centre at the Johns Hopkins University Applied Physics Laboratory in Laurel, Maryland, to celebrate the Pluto flyby.

PLUTO FACT FILE

- Pluto was named, not after the Disney dog, but by an English girl, Venetia Burney, who suggested the name of the Roman god of the underworld.
- A student-built instrument on NASA's New Horizons probe was named the Venetia Burney Student Dust Counter.
- New Horizons carries some of the ashes of its discoverer, Clyde Tombaugh, who died in 1997 at the age of 90.
- The IAU ruled that to be a planet, a world had to orbit the Sun, be big and massive enough to have collapsed into a sphere, and must have cleared its orbit of other objects.

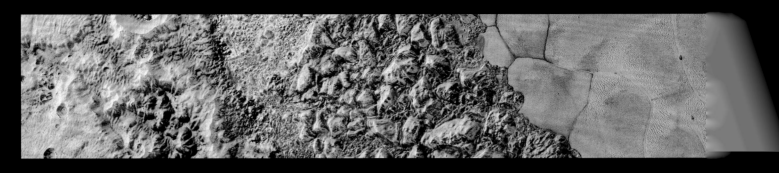

Other areas of Pluto are covered with ancient craters, and the contrast confirms Pluto to be an active world. Two mountains that stand several kilometres high, Wright Mons and Piccard Mons, have large holes at their summits and appear to be ice volcanoes. Other mountains are capped with methane snow. Amidst the jumbled terrain can be seen water icebergs floating in a sea of frozen nitrogen. Some of the surface resembles snakeskin, while peculiar ridges, dunes, glaciers and mottling indicate the interaction that is going on between water ice and other frozen gases. There may be an underground ocean of ammonia-rich water.

Another surprise was Pluto's atmosphere, a bluish haze that was discovered to contain many layers to an altitude higher than 200 km.

COMPANION CHARON

Pluto's largest moon is itself pretty sizeable, compared to its host world, and at 1,200 km across is about half its diameter. Less cratering than expected is again visible, but, intriguingly, Charon is covered with chasms and ridges including a series of canyons that run round its equator. They suggest there was an underground ocean that expanded as it froze, causing the surface to stretch and split open, possibly releasing a form of lava made of ice. Charon's outer layer is mainly water ice. Its north polar cap is reddened by a covering of a type of chemical compound called tholins.

▼ A view over Pluto's north pole shows long canyons, and pits where the ground has collapsed.

▲ Pluto's assorted terrain includes mountains, icebergs floating in nitrogen and smooth icy plains.

NEW TARGET 2014 MU$_{69}$

As New Horizons was on the final leg of its voyage to Pluto, astronomers were busy trying to find a new target for it to reach after the encounter. The Hubble Space Telescope was brought in for the search and discovered five objects that the spacecraft could reach with its available fuel. The mission team settled on one known as 2014 MU$_{69}$ that lies about 1.5 billion km beyond Pluto. Estimated to be about 45 km wide, it is thought to be one of the chunks left over when planets and worlds like Pluto were formed. New Horizons will reach it in January 2019.

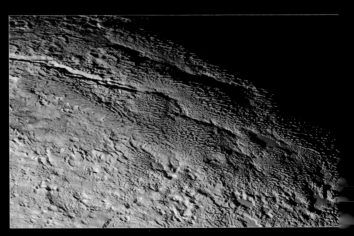

▲ A snakeskin-like region of rounded, blue-grey mountain ridges, with reddish material between them, is viewed close to Pluto's night side.

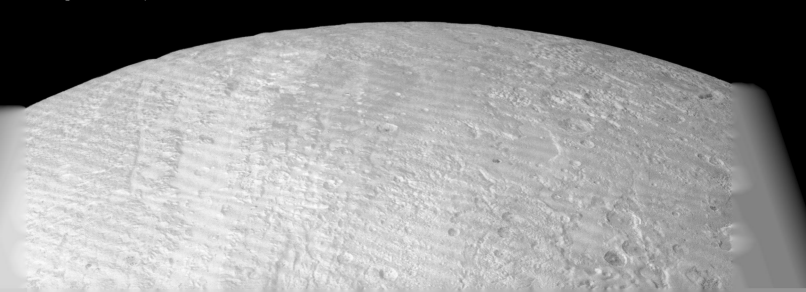

PLUTO'S ICY DOMAIN

Pluto's days as a mainstream planet were numbered when astronomers began to discover other worlds in the far reaches of the Solar System. Three have already been awarded dwarf planet status like Pluto, and others are sure to follow. Pluto, special as it is, clearly stands as gatekeeper to a disk-shaped zone that is packed with countless icy bodies.

THE KUIPER BELT

Though these remote objects are generally known as trans-Neptunian objects (TNOs), they are found in different groupings. The zone that Pluto inhabits was predicted to exist by astronomer Gerard Kuiper in 1951, like an icy version of the asteroid belt. It is known as the Kuiper Belt.

Our knowledge of this region is very much in its infancy. The first new object in it was found in 1992 and labelled 1992 QB_1. It is about 160 km wide. By early 2016, more than 1,300 Kuiper Belt objects (KBOs) had been detected. But these are just the tips of the celestial icebergs. When NASA wanted to find a fresh target for New Horizons following its Pluto encounter, the Hubble Space Telescope was able to discover five possibilities on its flight path within a few weeks!

US astronomer Mike Brown has become an authority on the Kuiper Belt, discovering several of the larger new worlds out there. He believes that dozens, if not hundreds, of KBOs could meet the criteria for dwarf planets. Many hundreds of thousands of icy objects bigger than 100 km are thought to inhabit the zone, and numerous short-period comets – those which swoop in on orbits of less than 200 years.

By 2016, the three KBOs accepted to be dwarf planets were Eris, Haumea and Makemake.

Eris: Found in 2005, this is the second largest dwarf planet, measuring 2,325 km in diameter. For a while it was thought it might be bigger than Pluto, but the discovery of a moon, Dysnomia, allowed scientists to gauge its size more precisely. Eris' steeply inclined and elongated orbit carries it around the Sun once every 561 years.

Haumea: This world, discovered in 2004, is odd for having a shape more like a rugby ball, measuring 1920 × 990 km. It orbits every 282 years, at an average distance of 5,795 billion km, rotates in only four hours, and has two satellites, Namaka and Hi'iaka. A mysterious red patch on its surface may be a sign of organic materials.

Makemake: Found in 2005, it orbits the Sun once every 310 years at an average distance of 6,783 billion km, and is about two-thirds the size of Pluto.

Other discoveries that are contenders for dwarf planets include the third largest found, currently known only as **2007 OR$_{10}$**, and 1,535 km wide; **Quaoar**, which is 1,170 km across, orbits once every 286 years at a distance of 6.5 billion km, and has a tiny moon, Weywot; **Orcus**, which is 983 km wide, orbits in 248 years, and has a satellite, Vanth; **Salacia**, 921 km across, which orbits every 271 years, and has a moon, Actaea; and **Varuna**, which takes 281 years to make one lap of the Solar System.

▼ An artist imagines how Sedna looks beyond the Kuiper Belt. Though a moon is shown, none has yet been discovered.

▲ From its remote location, the Sun appears as a brilliant, starlike point from the surface of Sedna, which takes 12,000 years to make one orbit.

THE OORT CLOUD

Far beyond the Kuiper Belt lies a much greater reservoir of billions of icy bodies, forming a ghostly shell around the whole of the Solar System. It has never been directly observed, but is named after Jan Oort, a Dutch astronomer who suggested its existence in 1950. The cloud is thought to begin about a light year from the Sun and may stretch a third of the way to the nearest star. It is a region filled with frozen fragments of water, ammonia and methane left over from the formation of the Solar System. The Oort Cloud is believed to be the source of so-called long-period comets. Though still theoretical, it neatly explains their random appearances.

▲ An imaginary view of the icy Kuiper Belt object Quaoar, one of the largest of the newly found occupants of this zone beyond Pluto.

REMARKABLE SEDNA

A group of icy bodies with far more extreme orbits exists outside the Kuiper Belt, and they are known as scattered-disk objects (SDOs). The most striking example is Sedna, whose distance from the Sun ranges from 11.5 billion km at closest to 145 billion km at its farthest. Sedna takes 12,000 years to complete one orbit!

▶ This depiction is of a smaller object in the Kuiper Belt, similar to 2014 MU₆₉, which NASA's New Horizons will visit in 2019.

COMETS – VISITORS FROM DEEP SPACE

As well as planets and asteroids, other bodies travel through our cosmic neighbourhood, occasionally becoming bright enough to appear spectacular. They are called comets and their appearance throughout history earned them a fearsome reputation. Today we know they are minor, insubstantial bodies compared to planets. But they are still regarded with awe by scientists because they are pristine chunks left over from the formation of the Solar System, 4.5 billion years ago. Analysis has shown that their chemistry is organic, including water ice, methane and carbon compounds beneath a dark crust. Our understanding of comets has been boosted by Europe's recent Rosetta mission to explore and land on one.

HAIRY, SCARY STARS

The word comet means 'hairy star', because the ones bright enough to be seen before the age of the telescope grew tails resembling flowing hair. They were unpredictable, unlike the planets in their steady courses, so the sudden addition of a bright interloper, with a tail stretching across the sky, unnerved court astrologers in early civilizations who saw them as portents of cataclysmic events.

The brightest comets are still unpredictable, because many are on their first journey into the Solar System. A spectacular recent example was Comet McNaught, which dazzled southern-hemisphere observers in 2007. Ten years earlier, Comet Hale–Bopp shone brightly for weeks. Astronomers now believe that most originate from a vast zone of icy debris called the Oort Cloud, stretching perhaps a third of the way to the nearest star. Something – such as a passing star long ago – disrupted the cloud, nudging some of this debris so that it fell in towards the Sun. Other comets, with shorter periods, are thought to be from the Kuiper Belt.

Comets typically have extremely stretched orbits that bring them in close to the Sun before sending them back out into space, sometimes never to return. But throughout history, Jupiter has wielded its gravitational influence to steer a number into much smaller orbits. These now orbit the Sun in periods of as little as a few years, and their appearance can be predicted.

Bright comets have always made the cosmic headlines. However, the vast majority are much fainter. Many are tracked across the sky every year by astronomers, but only a few become bright enough to be seen with binoculars, let alone the naked eye.

◀ The brilliant flash from Deep Impact's missile dazzles the space probe's camera after hitting Comet Tempel 1 in July 2005. Features recorded on the comet's nucleus included craters from more natural impacts.

WHAT IS A COMET?

The bundle of ice and dusty material that produces a comet is known as the nucleus and is usually no more than a few tens of kilometres wide. As it heads into the Solar System, it becomes warmed by the sunlight, thawing the ice and causing gases to escape and expand into a ghostly 'atmosphere' called a coma. Only the more active comets form tails, giving off a straight jet of gas that points away from the Sun due to pressure from the solar wind, and a second curved flow of dust released by the warmed nucleus. When the Earth passes through streams of dust left by comets, they produce showers of meteors, or 'shooting stars'.

HALLEY'S COMET

The classic example of a comet disturbed by Jupiter is the well-known Halley's Comet, which now travels into the inner Solar System from beyond Neptune every 76 years and was last close by in 1985–6. Comets usually take the names of their discoverers, but Halley's Comet was named after English astronomer Edmond Halley, who realized that certain comets recorded throughout history were one and the same, and so successfully predicted its return of 1758. We will next see it in 2061.

▶ Comet Hale-Bopp, a brilliant visitor in 1997, shows the classic twin tails, one formed of gas and a redder, curving flow of dust.

▼ One of the brightest examples of recent times was Comet McNaught, which became spectacular from the southern hemisphere, with a vast sweeping tail of dust.

VISITING COMETS

Though comets have visited us throughout history, the last return of Halley's Comet marked the first time space scientists went to visit them! Five spacecraft – two Soviet, two Japanese and one European – flew out to study it as it passed through the inner Solar System. ESA's Giotto flew closest, at a distance of just 596 km, and sent back the first images of the comet's nucleus, showing that it was shaped like a peanut.

In 2001, NASA's Deep Space 1 flew past Comet Borrelly – and in 2004, the NASA Stardust probe sped through the coma of Comet Wild 2, collecting particles that it sent back to Earth in a capsule. But NASA's most dramatic comet encounter came when a probe called Deep Impact fired a missile into Comet Tempel 1 on 4 July 2005. Scientists had wanted to know what lay within, but the explosion was so bright that the flash hid the view. Deep Impact, now renamed EPOXI, flew on to view Comet Hartley 2 in October 2010 – but without the fireworks this time.

ROSETTA – LANDING ON A COMET

Rosetta, the European Space Agency's new mission to a comet, lifted off in March 2004 from Kourou, French Guiana. The launch had been set back 14 months by the failure of an earlier Ariane 5 rocket, and the launch delay meant a new comet had to be found to replace its original intended target, 46P/Wirtanen.

The team settled on Comet 67P/Churyumov–Gerasimenko, named after its two Ukrainian discoverers, which was another comet with an orbit shortened by Jupiter. It now travels from beyond Jupiter to a closest point to the Sun, called perihelion, that lies between Mars and Earth. A single circuit takes six and a half years.

Originally, the mission had been designed to bring a comet sample back to Earth, but this was dismissed as too ambitious and the spacecraft carried a small companion probe, Philae, instead to land on the nucleus and study it.

Rosetta took a roundabout route to reach the comet, swinging past Earth and Mars to build up speed. On one such approach to Earth, the spacecraft was briefly mistaken

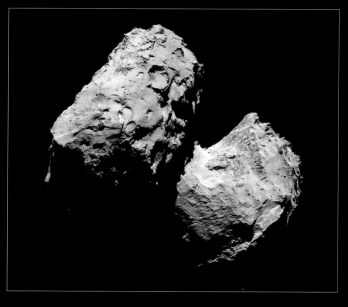

▲ A colour photo of Comet 67P/Churyumov–Gerasimenko reveals it to be actually very grey. The twin-lobed shape indicates that it was formed from two objects that collided.

for an incoming threatening asteroid! After visiting two asteroids *en route*, Rosetta was put into hibernation to conserve power in July 2011. Scientists were elated and relieved when it responded to a wake-up signal two and a half years later in January 2014.

Then, as the comet grew from being a tiny speck to show some form, the team realized they faced a new challenge. Rather than being round, it had a twin-lobed shape that resembled a duck. Landing Philae was going to be tricky already, since a comet has such a small gravitational pull, but it just got a lot more difficult. Rosetta rendezvoused with 67P in August 2014 and began to fly around it at varying distances in order to study it, but also to map its irregular form and decide where to send Philae.

◀ An artist's impression of the moment when Rosetta's companion probe, Philae, was released from the spacecraft to land on the head of the comet.

In November the probe was sent on its descent to touch down on the head of the 'duck'. But after harpoons failed to fire and anchor it securely, it bounced twice before ending up in the shadow of a cliff. Philae sent back much useful data for several hours before its batteries died and, without sunlight to recharge them, contact was effectively lost.

Meanwhile, Rosetta continued its work as the comet got closer to the Sun and began fizzing with activity, sending out jets of gas and dust. It reached perihelion in August 2015 before beginning the slow retreat into the outer Solar System again. Rosetta could have continued orbiting 67P, but its controllers decided to end the mission in style by landing Rosetta close to Philae on the comet in September 2016.

WHAT WE LEARNED

Traditionally, scientists have seen comets as 'dirty snowballs', but 67P showed their make-up is more complex. Comet 67P is certainly not shaped like a snowball, but appears to be formed from two objects that merged, billions of years ago.

The nucleus was found to be very low density, porous, and covered in sinkholes where the surface has collapsed as gas escaped to form the comet's coma and tail. Much of its surface is covered in powdery dust. There, and in the coma, have been detected organic compounds and other essential ingredients for life. But the water in the comet was of a different 'flavour' to that found in Earth's oceans, suggesting that our seawater may not have been delivered by the bombardment of comets like 67P.

A big surprise was the discovery of molecules of oxygen, which had not been expected to be able to survive so long in a comet – another way Rosetta is changing the way we understand how the Solar System formed.

▲ Rosetta's mission control team at Darmstadt, Germany, show their delight as the spacecraft awakes from years of hibernation, right on cue, in January 2014.

ROSETTA FACT FILE

- Rosetta was named after the Rosetta Stone, whose texts helped language experts unlock the meaning of Egyptian hieroglyphs.
- Philae was named after an island in the River Nile where an obelisk's description confirmed the interpretation of the Rosetta Stone's inscriptions.
- With its huge solar panels, Rosetta stretched 32 m in length. Philae was the size of a washing machine in comparison.
- Comet 67P is the size of a city, but its gravity is so weak that a tiny jump would allow an astronaut to escape its surface.

◄ The nucleus of Rosetta's comet, 67P, is overexposed to reveal the fainter jets of gas and dust being given off as it is warmed by the Sun on its approach to perihelion, forming an atmosphere-like coma and tail.

GAZING INTO THE UNIVERSE

Dusk falls over an engineering marvel of astronomy, the giant steerable Lovell Telescope radio dish at Jodrell Bank Observatory, in England. Throughout history, observatories have been following the heavens. Now advances in technology are allowing telescopes on Earth and in space to put the Universe under scrutiny as never before, including the discovery of countless planetary systems around other stars.

OBSERVATORIES ON EARTH

Humans have been curious about what was happening in the heavens since they gained the ability to think. There was no concept of astronomy as a science but it became clear that the positions of the Sun, Moon and stars at different times of the year marked important dates in the calendar, such as when to plant crops.

TEMPLES TO THE STARS

Early tribes built structures such as stone circles, of which Stonehenge in England is a famous example, and at least some of these might have had some astronomical as well as ritual purpose, for example marking a seasonal solstice or equinox. Similar edifices but with heavenly functions can be found in many civilizations, including Central America, China, India and the Middle East.

As societies developed and attempted to understand their surroundings, observatories were built around the world to explain the motions of celestial bodies. The telescope had yet to be invented, but other instruments were developed, such as the astrolabe, to measure the positions of stars and planets.

Tycho Brahe, a colourful Danish nobleman, made meticulous measurements in the 16th century that later allowed his assistant, the German mathematician Johannes Kepler, to formulate the laws that perfectly describe how planets, comets and other bodies orbit the Sun.

Generally, it was still accepted that the Earth was at the centre of events, though notably Aristarchus had proposed a system with the Sun at the centre as early as the 3rd century BC. It was not until the Renaissance when Nicolaus Copernicus presented the same idea and his reasoning for it that the idea began to find acceptance.

SEEING THE LIGHT

The telescope revolutionized astronomy. Though Italian astronomer Galileo Galilei did not invent it, he was one of the first to use it in the early 17th century to observe the Moon and planets, along with others including Thomas Harriott in England. Crude by modern standards, these were refracting telescopes that used lenses to collect and magnify the light received. Nevertheless, they allowed users to see craters on the Moon, the four main satellites of Jupiter and the rings of Saturn.

Isaac Newton, the famous English scientist who described the nature of gravity and explained some of the most important laws of physics, invented a different type of telescope, one that

▲▶ An image taken with a backyard telescope shows the galaxy M51 in exquisite detail above, but the Third Earl of Rosse's sketch in 1845 (right) first revealed its spiral structure

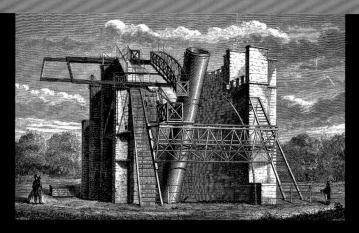

▲ A contemporary engraving shows the mighty 'Leviathan' telescope in Ireland that was once the world's largest.

▶ The historic Royal Observatory Greenwich, which is home to the Prime Meridian.

used a curved mirror rather than a main objective lens to collect light. His design for the reflecting telescope has been adapted in various ways but remains the basic model for today's powerful professional optical observatories, including the Hubble Space Telescope.

Important observatories were set up around the world following the invention of the telescope, such as the Paris Observatory in France and the Royal Greenwich Observatory in London, England, in the late 17th century. Their large, mounted telescopes were primarily used in the early days to measure precise positions of the stars rather than to conduct research, in order to help sailors to determine their position at sea.

While it was straightforward to ascertain one's latitude, longitude was a different matter because the Earth was turning. Accurate positions together with more reliable clocks allowed seafarers to tell how far they were from a fixed, but arbitrary, line of longitude called the Prime Meridian. With typical rivalry, Paris and London each set their own Prime

Meridian, running through their respective observatory, but the world eventually settled on London as the point from which all longitudinal positions are determined.

Observatories were focusing more on astronomical research by the 19th century, using telescopes to learn more about the Sun and objects in the night sky. Telescopes were being made with high-quality optics, and could be used with a newly designed instrument called a spectroscope to examine starlight and see what the star was made of.

Observatories began to spring up around the world, often built by wealthy individuals. From 1845 to 1917, the world's largest telescope was a giant reflector dubbed the Leviathan of Parsonstown, based on Isaac Newton's classic design, and built by the Third Earl of Rosse at Birr Castle in Ireland. With its 72-inch (1.8 m) mirror, it was the first to reveal a spiral pattern in certain fuzzy blobs in the night sky, indicating their true nature as separate galaxies rather than nebulous gas clouds within our own Milky Way.

▼ Comet Hale–Bopp shines as night falls over Stonehenge in Wiltshire, a prehistoric stone circle whose purpose may have included recording seasonal changes in the heavens.

MODERN ASTRONOMY

Most early astronomical telescopes were installed in observatories in city centres or similarly convenient locations. But by the end of the 19th century, it was being realized that better observing conditions could be achieved by placing telescopes high on hills or mountains, above much of the air currents in the densest part of the atmosphere. The growth of street lighting was harming attempts to catch faint starlight and remote mountains offered another advantage here.

Placing a telescope on a mountain could be a very challenging operation, but the determination and enterprise of the day saw such observatories as the Pic du Midi, built in the late 19th century, which looks almost impossibly perched on a summit in the French Pyrenees.

Soon after, work began on building a similar observatory on Mount Wilson, not far from Los Angeles in California. In 1917, it became home to the 100-inch (2.5 m) Hooker Telescope that was the world's biggest and allowed leading astronomer Edwin Hubble to show that our Milky Way was just one galaxy among many in an expanding Universe. The telescope, whose mirror has a similar size to the space telescope named after Hubble, remained the largest until the 200-inch (5.1 m) Hale Telescope opened on Mount Palomar in 1949.

In the latter part of the 20th century, with the growing problem of light pollution from cities such as Los Angeles affecting astronomers' work, new observatories were designed for some of the world's highest and most remote locations. They include the summit of Mauna Kea on Hawaii, La Palma

▲ The large steerable dish of the Green Bank Telescope in West Virginia, with a smaller ancillary instrument in the foreground.

in the Spanish Canary Islands, which hosts the world's biggest optical instrument, the Gran Telescopio Canarias, and the mountains of northern Chile. As well as being protected from artificial light, the sites also often enjoy steady atmospheric conditions thanks to the prevailing winds bringing undisturbed air in from over the sea. Astronomers rarely put their eye to such telescopes nowadays, and they are fitted with complex instrumentation instead.

► Throughout the second decade of the 21st century, the biggest optical instrument in the world has been the Gran Telescopio Canarias on La Palma, with its 10.4 m mirror.

▲ The dishes of ALMA, the highest observatory on Earth and arranged on a plateau in Chile, are carrying out cutting-edge astronomical research.

RADIO TELESCOPES

Until the 20th century, astronomical observations were all carried out by studying visible light. But in the early 1930s, interference to shortwave radio broadcasts was investigated, and its timing showed that it was coming from the heart of our Milky Way. The science of radio astronomy was born. Early radio telescopes resembled hop poles while others took the form of small parabolic dishes, reflecting incoming radio waves just as curved mirrors had collected starlight. It was found that the Sun and Jupiter were both powerful emitters of radio waves.

Large radio telescopes were developed, including the 100 m diameter steerable dishes of the Green Bank Telescope in West Virginia, USA, and Effelsberg, near Bonn, Germany, and the 76 m Lovell Telescope at Jodrell Bank, near Manchester, England. The world's largest fixed dish, 303 m wide, sits in a valley at Arecibo in Puerto Rico, but a larger rival with a 500 m dish was nearing completion in 2016 in the Guizhou Province of south-west China.

Due to the long wavelength of radio waves, individual radio telescopes are unable to observe objects in great detail. However, it was found that telescopes could be combined, either in a local array, or spread much further afield, to improve resolution. Together they mimic the power of a telescope as large as the distance between them.

The highest observatory in the world is a special type of radio telescope called ALMA (the Atacama Large Millimetre/submillimetre Array) at an altitude of more than 5,000 m in Chile's Atacama Desert. It uses 66 dishes to observe millimetre waves from cold, dark dust clouds that would be invisible to a normal telescope. This helps astronomers learn such things as how stars and galaxies formed, and how new planetary systems are created.

▲ An early mountain-top observatory, at the Pic du Midi in the French Pyrenees, which offered a view of the heavens away from city lights. It currently houses the largest telescope in France.

TELESCOPES OF THE FUTURE

THE SQUARE KILOMETRE ARRAY

A powerful new telescope called the Square Kilometre Array (SKA) is being built across South Africa and Australia that will use the technique of combining a number of telescopes together to observe the Universe. When completed, its many thousands of dishes and antennas will be sensitive enough to detect an airport radar on a planet tens of light years away. They will allow it to probe the very early Universe by imaging the formation of the first galaxies to shine. The telescope will have the processing power of 100 million PCs, and the optical fibre needed to connect all the instruments will be long enough to wrap around the Earth twice. Though the telescope itself will be in the southern hemisphere, its headquarters is at Jodrell Bank in the UK.

THE GIANT MAGELLAN TELESCOPE

A giant optical observatory being planned by 11 international partners is the Giant Magellan Telescope (GMT), which will be built at Las Campanas in northern Chile. It will collect light with seven 8.4 m wide mirrors that will give it a combined diameter of about 25 m, allowing it to see the Universe in up to ten times more detail than the Hubble Space Telescope. The GMT is due to be completed by 2021, when it will see 'first light' by making its first observations, and be fully operational by 2024.

▼ The European Extremely Large Telescope as it will look when it is completed on the Cerro Armazones peak in Chile.

▲ How some of the dishes that will make up the Square Kilometre Array will look when spread across the plains of South Africa.

THE EUROPEAN EXTREMELY LARGE TELESCOPE

European astronomers, who already have a number of telescopes in the ideal conditions of Chile, have commenced work building a telescope that will be the biggest optical instrument in the world. The name is rather unimaginative – the European Extremely Large Telescope (E-ELT) – but it will be 12 times more powerful than the current largest telescope on Earth, collecting 12 times more light with its 39.3 m wide

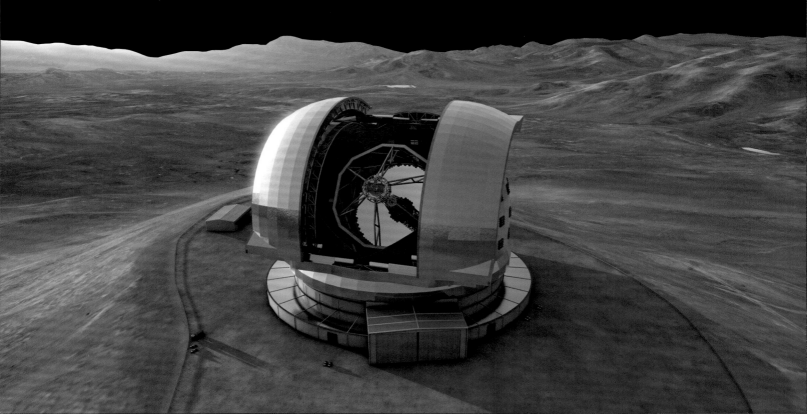

▲ An artist's impression of the Giant Magellan Telescope, which is due to open in Chile in 2021.

segmented mirror. It will stand on a mountain called Cerro Armazones, which has already had its 3,000 m high peak levelled by explosives. The telescope will sit in an area the size of a sports stadium, observing in visible and infrared light. It will allow astronomers to peer back billions of years in time to view the first galaxies, and directly image exoplanets around nearby stars, helping to identify rocky worlds like Earth.

THE LARGE SYNOPTIC SURVEY TELESCOPE

The Large Synoptic Survey Telescope (LSST) is another instrument being planned for Chile, which will become an ultra-powerful, 3-billion-pixel, digital camera on the sky. It is planned to start operating in 2021 on the Cerro Pachón ridge, and over ten years its 8.4 m primary mirror will act like a giant camera lens to take shots that each cover nearly 50 times the apparent area of the Moon. By continually monitoring the sky, real-time changes can be observed, allowing this telescope to identify near-Earth asteroids, as well as tackling mysteries of the deeper Universe such as dark energy and dark matter, and the structure of the Milky Way. Each night the telescope will generate more than 30 terabytes of data.

▶ An artist's conception of the Large Synoptic Survey Telescope which will act as a giant digital camera on the night sky.

A RADIO TELESCOPE ON THE MOON?

A future giant new radio telescope could be placed on the far side of the Moon, which is always turned away from the Earth. Radio astronomers are keen because the location would be shielded from all the radio noise generated by radio and TV stations, microwave ovens, radar, mobile phones, or reflected by the Earth's ionosphere. Such a telescope could detect extremely faint signals from the neutral hydrogen that filled the early Universe before any stars or galaxies began to form.

▲ The shape of things to come? A lunar base is set up to support a radio telescope, seen fixed in the lunar soil, to probe the Universe away from interference.

NEW WINDOWS ON THE UNIVERSE

As well as developing probes to explore the Moon and planets, the Space Age opened up new windows on the Universe for astronomers. Rocket technology allowed them to put telescopes far above the Earth's atmosphere. There they could operate free from disturbing air currents, and also observe in regions of the light spectrum blocked by the atmosphere. Some space telescopes are orbiting Earth, while others are at the Lagrangian points where a satellite can sit continuously in harmony with both the Earth and the Sun.

▲ Dubbed the Pillars of Creation, this image of giant columns of cold gas lit by young stars was taken by the Hubble Space Telescope.

THE ELECTROMAGNETIC SPECTRUM

William Herschel became famous for discovering the planet Uranus. But he also made an important discovery about light. Most people know that a prism will break up white light into an array of colours from red to violet, called a spectrum, and raindrops do the same in nature to produce a rainbow from sunlight. Herschel found that this spectrum extends beyond what is visible. When he placed a thermometer beyond the red end of the spectrum, it showed a rise in temperature. Herschel had found infrared radiation. Today we know that visible light is just a small section, which our eyes are attuned to, within a much greater band of frequencies that is called the electromagnetic spectrum. It stretches from radio waves, which have the lowest energy and long wavelengths, to gamma-rays, which are highly energetic and of very short wavelength.

▼ The glowing echo of the early Universe known as the cosmic microwave background is mapped by ESA's Planck satellite, helping to peg its age at 13.81 billion years.

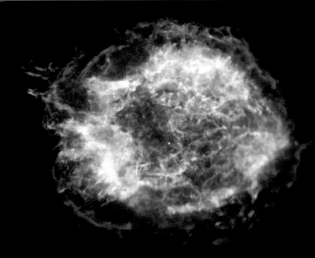

▲ The remnant of an exploding star, or supernova, called Cassiopeia A is pictured in high-frequency light by NASA's Chandra X-ray Observatory.

Radio: The same radiation that is emitted by radio stations is broadcast naturally by stars and galaxies. Some galaxies shine brightly in radio waves, as well as quasars and pulsars, the Sun and Jupiter.

Microwave: Radiation of the kind used to heat your ready-meal is being given off by the Universe. Astronomers use it to study galaxy structure and the echo of the Big Bang.

Infrared: This form of radiation allows astronomers to measure the heat of the Universe. It is absorbed by the Earth's atmosphere, so in order to study it observatories had to be placed high on mountains, before they had the solution of putting 'thermal imaging' telescopes in space. Infrared telescopes help scientists map otherwise invisible dust between stars, and to peer through it to study the heart of our Milky Way Galaxy.

Visible: The form of electromagnetic radiation with which most of us are familiar because we can study it with our own eyes. The Hubble Space Telescope is a famous example of a telescope observing in visible light (as well as other frequencies) in space.

Ultraviolet: Another type of light that is strongly filtered by our atmosphere and so must be observed from space. It is particularly useful for studying the hotter stars as well as the interstellar medium of gas and dust between stars. UV radiation from the Sun is what causes sunburn.

X-ray: An important field of space research, because the same radiation that allows a dentist to check your teeth, or airport security to examine your bags, is also given off by extremely hot sources in space. They include interacting binary stars, gas in clusters of galaxies, supernova remnants and the centres of galaxies where black holes lurk.

Gamma-ray: Used for medical imaging, this high-frequency radiation is given off by some of the most powerful objects in the Universe, including pulsars, material falling into black holes and intense solar flares. Incredibly powerful gamma-ray bursts are routinely detected, with a million times more energy than given off by a whole galaxy, They are thought to be caused by the collapse of the most massive stars, or the merger of pairs of neutron stars or black holes.

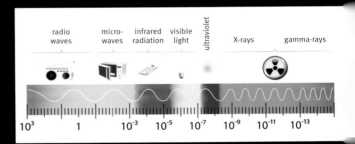

▲ Different parts of the electromagnetic spectrum beyond the rainbow of visible light are shown in this illustration.

▲ This unusual view of the closest spiral galaxy to our own Milky Way, the Andromeda Galaxy M31, was taken by NASA's Swift satellite in ultraviolet light.

▲ An artist's impression of the Herschel Space Observatory set against an image it took in infrared light of baby stars forming in the Rosette Nebula, 5,000 light years away.

THE HUBBLE SPACE TELESCOPE

There have been many observatories launched into orbit, but none has had as much impact as the Hubble Space Telescope that is operated jointly by NASA and ESA. This moderate-sized professional instrument, operating in ideal conditions, has made countless discoveries and performed invaluable research. But more than that, it has brought the excitement and awe of the Universe into people's lives thanks to a stream of spectacular images of stars, planets, gas clouds and distant galaxies.

It had long been realized that the vacuum of space would be a perfect place from which to do astronomy, and plans for such an instrument were being discussed by American astronomers in 1946. Hubble survived attempts by politicians to cancel it, then its launch was delayed by the *Challenger* disaster that grounded the Space Shuttle fleet. But finally it was lifted into orbit by *Discovery* in April 1990, to be released from the Shuttle's cargo bay, 560 km above the Earth.

The telescope – named after Edwin Hubble, who confirmed that the Universe was an expanding collection of galaxies – is a reflector, of the basic type invented by Sir Isaac Newton. Its Cassegrain design is similar to that used by many amateur astronomers with their commercial backyard telescopes, but on a much bigger scale. The telescope is 13.3 m in length and has a primary mirror 2.4 m wide. Solar panels provide its power and special sensors lock on to guide stars to make sure it is pointed precisely at its observing targets.

▲ Astronauts Michael Foale, left, and Claude Nicollier, on Space Shuttle *Discovery*'s robotic arm, install a new instrument on the Hubble Space Telescope during a servicing mission in 1999.

When Hubble achieved 'first light' – the term for when it makes its initial observations – it became clear that there was a serious problem with the telescope. A small but important error in grinding the curve of its main mirror meant that the images it was taking were distorted.

Fortunately, because Hubble had been placed in an accessible orbit around the Earth, NASA was able to come up with a solution. Replacing the main mirror was out of the question, but engineers came up with a correcting lens that could be placed in the telescope's optical path, to act rather like a pair of spectacles. Astronauts on the Shuttle *Endeavour* fitted the device on the first mission to service Hubble in 1993 and it repaired the telescope's 'eyesight' perfectly.

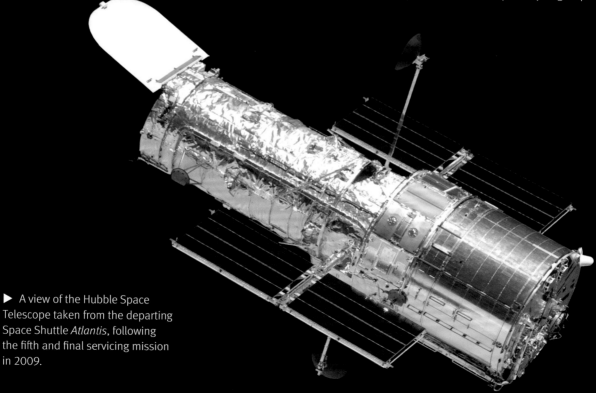

▶ A view of the Hubble Space Telescope taken from the departing Space Shuttle *Atlantis*, following the fifth and final servicing mission in 2009.

PINPOINT PRECISION

Hubble is so steady it can lock on to a target without deviating more than a hair's width when seen from 1.5 km away. Its main mirror's surface is so smooth that if it was scaled up to the diameter of the Earth, the biggest bump would be just 15 cm high.

Though famous for its photographic studies of all kinds of objects, Hubble does not just operate in the visual part of the spectrum. One of its two cameras, and three other instruments, also view the Universe in ultraviolet, visible and infrared light.

Hubble has carried out some memorable astronomy. Its imaging of remote galaxies has helped astronomers to gauge the age of the Universe and to measure how fast it is expanding. It has taken long-exposure images of tiny, apparently empty, patches of sky and shown them each to contain thousands of galaxies. In our own Milky Way, it has taken detailed images of gas clouds that are the birthplace of stars, as well as the debris left by dying stars as supernovae. It took the first photo of an exoplanet, in the dust around the star Fomalhaut, and identified gas in other exoplanets' atmospheres. And in our own Solar System, Hubble has taken high-quality photos of the planets, including weather on Mars and a spectacular comet impact on Jupiter. It discovered moons of Pluto and potential targets in the Kuiper Belt for the New Horizons probe to visit, as well as measuring other icy bodies in that distant zone.

Following the fitting of Hubble's 'spectacles', there were four more Shuttle missions to service the telescope, update its instruments and keep it functioning. But no further expeditions are planned and its days now seem numbered. It is likely to be sent to a fiery end in the Earth's atmosphere some time in the 2020s.

▶▼ Hubble has produced many amazing images, including a tower of gas and dust (right) in the Eagle Nebula, and below (from left to right) a light echo from dust around star V838 Monocerotis, a butterfly of streaming gas from a dying star, a blue bubble blown by a massive star, and a stellar nursery called the Cone Nebula.

HUNTING NEW PLANETARY SYSTEMS

A major advance in astronomy in the last few years has been the discovery of hundreds of planets orbiting other stars. Termed exoplanets, they have given themselves away thanks to the use of more sensitive detectors on ground-based telescopes as well as specialist observatories in space.

The first exoplanets were found orbiting a pulsar in 1992. But in 1995, two astronomers discovered the first planet orbiting an ordinary star – numbered 51 in the constellation of Pegasus. Since then, more than a thousand exoplanets have been confirmed and many others are suspected.

Early discoveries were made using what is known as the radial velocity method – instruments attached to big telescopes on Earth measured the wobble of a star that indicated it was being orbited by another object. But a further way of finding exoplanets was discovered using a technique called the transit method. This is where a planet reveals its presence by passing in front of its home star, causing a small but measurable fade in the star's brightness.

▲ A representation of NASA's Kepler Space Telescope, on its new K2 mission to discover more exoplanets around red dwarf stars.

KEPLER SPACE TELESCOPE

A number of observatories around the world started to look for exoplanets using the transit method. But the breakthrough that allowed hundreds to be discovered was a NASA space telescope called Kepler.

The Kepler Space Telescope was launched in March 2009. It has a 1.4 m mirror to collect light and was designed to stare constantly at a small area of sky within the constellations of Lyra, Cygnus and Draco, watching hundreds of stars at a time for the tiny fades that would give away an exoplanet.

Kepler has been amazingly successful, identifying more than 1,000 exoplanets. But in 2012, one of four stabilizing parts of the spacecraft, called reaction wheels and designed to keep it pointing correctly in space, failed, followed by a second wheel in 2013. With only two remaining, the telescope was no longer able to perform its original mission. However, the engineers came up with an ingenious solution to allow it to carry on hunting for exoplanets. This involved orienting the spacecraft so that the force of sunlight on it would act as a stabilizing influence – introducing a virtual reaction wheel. To make this work, the telescope had to be angled so that

▼ An artist imagines how an Earth-sized planet discovered by the Kepler Space Telescope might look, orbiting in the habitable zone of its home star.

▼ A selection of habitable zone planets discovered by Kepler, showing their estimated relative sizes to the Earth. No one really yet knows how they look.

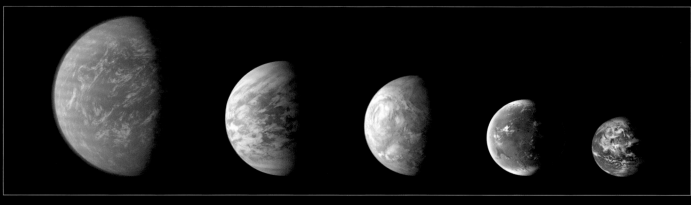

▲ This world, Kepler-452b, is about 60% bigger than Earth and was found 1,400 light years away in the habitable zone of a star that is similar to our Sun. Scientists do not know if it can actually support life.

◀ How NASA's next planet-hunting telescope, TESS, will look as it scours deep space. By monitoring more than 200,000 stars, it should add hundreds more worlds to those already known.

it was pointing in the plane of the Earth's orbit around the Sun, and it would have to turn to observe a fresh patch every 80 days or so to compensate for the Earth's own orbital motion, to stop sunlight getting into the telescope.

The workaround was very successful and Kepler began discovering exoplanets again, around red dwarf stars, on its new mission, labelled K2.

TYPES OF EXOPLANET

The earliest exoplanet discoveries showed that other planetary systems are quite unlike our own Solar System. Many of the discoveries have been giant gasball planets orbiting close to their host stars – so-called 'hot Jupiters'. Several stars have been found to have more than one planet, and a growing number of these are being identified as rocky worlds. The holy grail for exoplanet hunters is to find a rocky planet within the star's habitable zone – the region where liquid water could exist on a planet's surface. So far, a handful of contenders have been identified, generally a little larger than our own planet and so dubbed 'super-Earths'.

PLANET HUNTERS OF THE FUTURE

New space telescopes are being planned to boost the discovery of exoplanets. NASA's next planet hunter will be the Transiting Exoplanet Survey Satellite (TESS), which will survey the entire sky over time to monitor the brightness of more than 200,000 stars and watch out for short-lived fades caused by planetary transits. Due to launch atop a SpaceX Falcon 9 in 2017, the telescope will operate in a high-Earth orbit and provide discoveries for the upcoming James Webb Space Telescope (see page 174) to follow up.

Europe is designing its own spacecraft, the Planetary Transits and Oscillations of stars mission (PLATO). It will use 34 small telescopes and cameras to watch up to a million stars for dips in brightness, in a bid to find rocky worlds. The satellite's other goal is to detect 'starquakes' to learn more about the host stars as well as the planets. PLATO is scheduled to be launched by a Soyuz rocket from Kourou, French Guiana, by 2024.

OTHER SPACE TELESCOPES

RadioAstron is a radio telescope launched by Russia in 2011 into a high orbit that carries it out as far as the Moon. Also known as Spektr-R, it has a 10 m dish to collect signals and is most effectively used together with radio observatories on the ground to act like a single telescope bigger than the Earth. The technique, called interferometry, allows radio astronomy imaging to be carried out in high resolution.

Europe's **Herschel Space Observatory**, with its 3.5 m mirror, made extensive observations of the infrared sky, following its launch from French Guiana in 2009, that will keep astronomers busy for years. It operated for four years at a point 1.5 million km from Earth, until the coolant needed to shield the telescope from its own interfering heat ran out.

Planck was a companion space telescope launched with Herschel to the remote Lagrangian 2 point in space. There it studied the afterglow of the Big Bang, called the cosmic microwave background, until it ran out of coolant and was switched off in 2013. A major finding was that the first stars in the Universe formed more than 100 million years later than had been thought.

NASA's **Spitzer Space Telescope**, launched in 2003 from Florida, was designed to make infrared observations for a range of targets, from asteroids in the Solar System to planet-forming disks, exoplanets and distant galaxies. Though now operating in a warm state, it is still functioning and has recently mapped the temperatures on the surface of an exoplanet 40 light years way.

▲ How ESA's Herschel Space Observatory might have looked as it carried out thermal imaging of the sky from 2009 to 2013, studying infrared light.

Another NASA space telescope, the **Wide-field Infrared Survey Explorer** (WISE), launched from Vandenberg, California, in 2009, to survey the entire sky. After its coolant ran out, the mission was reinvented as NEOWISE to search for near-Earth asteroids that could threaten a collision.

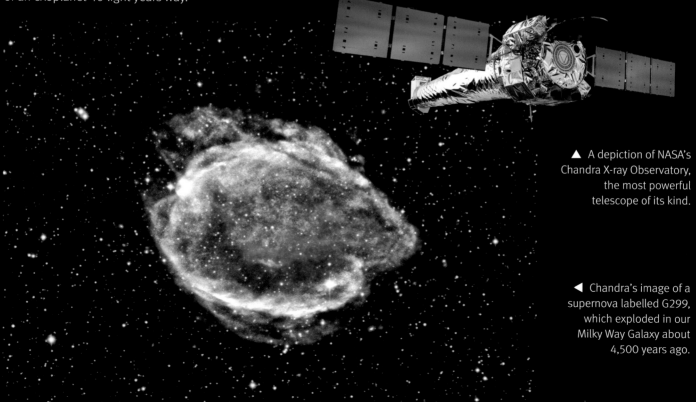

▲ A depiction of NASA's Chandra X-ray Observatory, the most powerful telescope of its kind.

◄ Chandra's image of a supernova labelled G299, which exploded in our Milky Way Galaxy about 4,500 years ago.

Akari was a Japanese space telescope launched in 2006 that spent more than five years scanning almost the whole sky in infrared light before suffering electrical failure in 2001.

NASA's **Galaxy Evolution Explorer** (GALEX) was an ultraviolet telescope launched from Florida in 2003 that spent ten years studying star formation and discovering new galaxies.

Chandra X-ray Observatory was launched from the cargo bay of Space Shuttle *Columbia* in 1999, into an orbit that takes it a third of the way to the Moon. A system of nested mirrors and a powerful imager allow this NASA telescope to observe X-ray light 100 times fainter and with much sharper resolution than any telescope before it from high-energy regions of the Universe, such as black holes and supernovae.

In 2012, NASA launched a smaller satellite, the **Nuclear Spectroscopic Telescope Array** (NuSTAR), to observe higher-energy events than Chandra and an ESA X-ray telescope, XMM-Newton. NuSTAR's achievements have included measuring the spin of a supermassive black hole and mapping material within a supernova remnant.

The **Compton Gamma Ray Observatory** was the second of NASA's 'Great Observatories' to be sent into space, following Hubble. It was placed in orbit by the Space Shuttle *Atlantis* in 1991 and worked for nine years before a gyroscope failed and it was brought back into the atmosphere and destroyed.

▶▼ ESA's Gaia satellite is making a detailed map of the Milky Way (illustrated right), while NASA's WISE space telescope performs a new mission to search for threatening asteroids (illustrated below).

Swift is a NASA mission, supported by the UK and Italy, that responds rapidly to gamma-ray bursts from colliding neutron stars, alerting astronomers on the ground so that other telescopes may be turned quickly on to them. It was launched in 2004 from Florida.

NASA's **Fermi Gamma-ray Space Telescope** observes the most energetic phenomena in the Universe including gamma-ray bursts, supermassive black holes and streams of hot gas travelling close to the speed of light. Launched in 2008, the spacecraft maps the whole sky every three hours.

Gaia is a European observatory, launched from French Guiana in 2013 to the L_2 point – it is making the most detailed 3D map of our Milky Way Galaxy ever by measuring the motions of a billion stars. It is also discovering asteroids, quasars and 'failed stars' called brown dwarfs.

LISA Pathfinder is a European spacecraft testing technologies for a later observatory that will try to detect ripples in space-time called gravitational waves. An upcoming ESA mission, **Euclid**, will investigate the nature of dark energy and dark matter by accurately measuring how the expansion of the Universe is accelerating.

NASA'S NEXT BIG TELESCOPES

THE JAMES WEBB SPACE TELESCOPE

The successor to Hubble will be a much bigger telescope, so powerful that it will be able to see planets around nearby stars and galaxies almost as far back as the Big Bang. Known as the James Webb Space Telescope after an early NASA administrator, the telescope is due to be launched in late 2018 by an Ariane 5 rocket from French Guiana. It is an international collaboration between the United States, Canada and Europe.

The telescope, known as Webb for short, will have a mirror 6.5 m wide, compared to Hubble's 2.4 m mirror, and it is being assembled in 18 segments, made of ultra-lightweight

▲ An engineer is reflected in a segment of Webb's primary mirror as it undergoes cryogenic testing to prepare for space.

beryllium, to resemble a honeycomb. Unlike Hubble, with its closed tube, Webb will have an open design. The whole telescope will be shaded from the Sun's heat by a five-layered sunshield the size of a tennis court. The assembly is so big that it will have to be folded up during launch and will open out automatically, to within an accuracy of a millimetre, when it is sent into space.

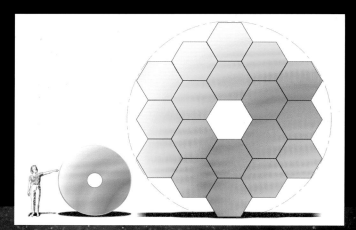

◄▼ A comparison (left) between the size of Webb's segmented mirror and the single smaller mirror that is part of the Hubble Space Telescope. Below is an artist's impression of how Webb will look when it is operating in deep space.

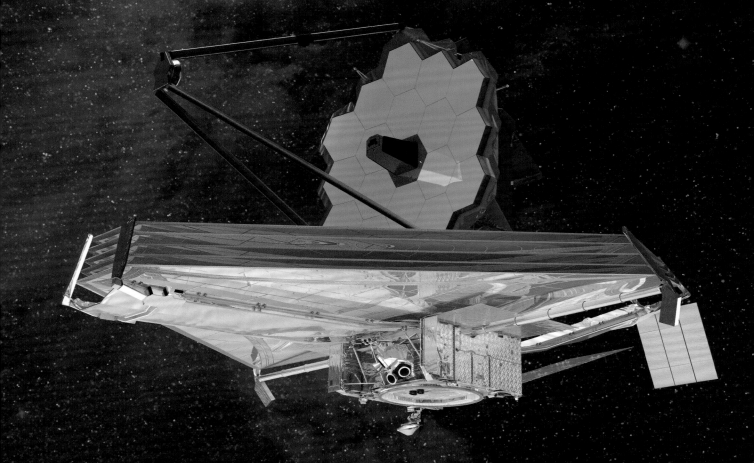

▲ Engineers work on a scaled model of Webb to check how its systems can test for any errors in the optics.

This time it is essential that nothing goes wrong, because there will be no opportunity to service Webb or to correct any faults. The telescope is being sent to one of the Lagrangian points, L_2, where it can observe in a gravitationally stable spot relative to Earth and the Sun – but that spot is out of reach of astronauts at a distance of 1.5 million km.

Webb's greater size will allow it to collect more than five and a half times as much light as Hubble, focusing on the infrared Universe rather than visible light. Four instruments, a combination of cameras and spectrometers, will be able to record incredibly faint objects. A spectroscope called NIRSpec will be able to observe 100 objects at once. Another, MIRI, will record the faintest galaxies as well as cold comets and Kuiper Belt objects. Both instruments are being contributed by ESA.

An American instrument, NIRCam, will detect faint objects that lie close to bright neighbours in the sky by blotting out the light of the more luminous objects. While gyroscopes and star trackers will help orient the telescope, a Canadian imager, known as FGS-TFI, will ensure it is pointed precisely in the right direction to take its detailed pictures.

Webb is designed to operate for at least five and a half years, and hopefully more than ten. The use of a sunshade to keep it cold should allow it to perform for far longer than NASA's Spitzer and ESA's Herschel telescopes, which both relied on a limited supply of coolant.

ATLAST – AN EVEN BIGGER TELESCOPE

With Webb not even launched, NASA is already thinking ahead to an even bigger and more powerful telescope to succeed it. It currently has the unwieldy name of the Advanced Technology Large Aperture Space Telescope (ATLAST) and it should be launched some time between 2025 and 2035.

A detailed design is not yet agreed, but the telescope would have a giant light-collecting mirror, 20 m in diameter, and segmented like Webb's. It will be so big that the telescope will probably need to be assembled by astronauts in orbit before it flies to its final observing location. But its power will allow astronomers to observe exoplanets with relative ease and to study their atmospheres for signs of life, such as oxygen, water and methane.

Like Hubble, ATLAST will observe in visible light as well as at ultraviolet and infrared wavelengths, but at significantly higher resolution.

▶ One possible design for the Advanced Technology Large Aperture Space Telescope (ATLAST) that is being considered by NASA's Goddard Space Flight Center for a future generation observatory that will be powerful enough to observe planets around nearby stars.

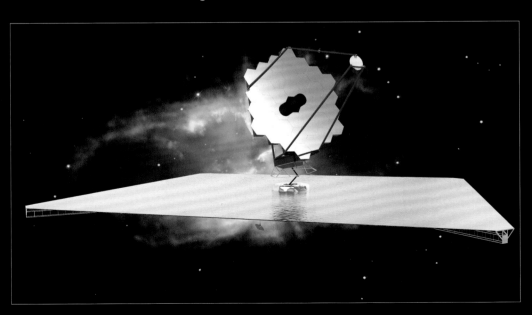

FUTURE SPACE EXPLORATION

Up until now, our space probes have only ventured as far as the outer Solar System. Other stars' planets — such as these pictured by an artist, circling a red dwarf 130 light years away — remain out of reach because of their sheer distance. But as a new age of space tourism dawns, crewed missions to set up colonies on Mars and beyond are also being planned, as well as concepts to make interstellar travel a reality.

FUTURE OF HUMANS IN SPACE

Years after the retirement of the Space Shuttle, NASA is building its biggest and most powerful rocket yet, with the capability to send astronauts out into the Solar System. Known as the Space Launch System (SLS), the rocket will come in a variety of configurations to suit different missions, including sending robotic probes to distant planets. Its crewed versions will launch NASA's Orion Multi-Purpose Crew Vehicle, a roomier and more comfortable version of the Apollo Command Module, which will fly astronauts to the Moon, asteroids and eventually Mars.

MANNED MISSIONS OF THE FUTURE

Though resembling the Saturn V rockets that launched Apollo to the Moon, the SLS rocket's central core stage, filled with liquid fuel, will have a pair of solid-fuel boosters attached in a development of the design used for the Shuttle. They will launch from the Kennedy Space Center, Florida, like the Saturns and Space Shuttles before them.

The first SLS mission with an unmanned Orion spacecraft – Exploration Mission-1 (EM-1) – will launch in 2018, flying beyond the Moon before returning the capsule to Earth, in a demonstration of the technology. It will be followed by a mission (EM-2) carrying up to four astronauts to the vicinity of the Moon. The SLS configuration will be the smallest, known as SLS Block 1. A more powerful model, known as SLS Block 1B, and taller than the Saturn V, will be developed for more ambitious crewed missions deeper into space.

Two other configurations, SLS Block 1B Cargo and SLS Block 2 Cargo, will have the Orion spacecraft replaced by fairings of different sizes to launch satellites, robotic probes and other hardware. When humans go to Mars, the larger model will travel on ahead carrying the materials that the first crews will need to set up a base on the planet.

▼ Astronauts enter a small lunar outpost as humans return to the Moon, in this concept artwork from NASA.

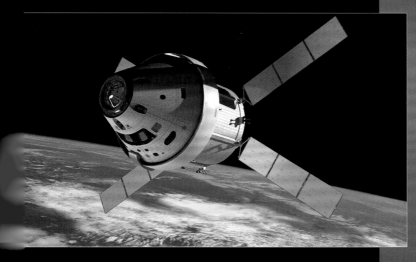

An artist's impression of NASA's Orion spacecraft in orbit, mated to its European-built service module and ready to embark on a crewed mission into deep space.

How a lift-off will look as NASA's new Space Launch System thunders into the sky from Florida, carrying astronauts beyond Earth orbit to the Moon, asteroids or Mars.

THE ORION MULTI-PURPOSE CREW VEHICLE

Orion's resemblance to Apollo is clear, but the crew module is large enough to carry six astronauts. There are two other main sections: a Service Module that was developed by the European Space Agency (ESA), and a Launch Abort System that can fly the crew out of danger in the event of an emergency, and which is jettisoned when Orion reaches Earth orbit. The Orion had a successful uncrewed test flight in December 2014, making two orbits to test its systems, following launch by a Delta IV Heavy rocket. Like Apollo, it then splashed down in the Pacific Ocean.

RIVAL ROCKETS

A new family of rockets called Angara has been developed for Russia's space agency, Roscosmos, to replace the veteran Proton, including a heavy-lift version, Angara 5, which made its first test flight in December 2014. The option for even more powerful models has been put on the back burner, and so adaptations of the Angara 5 are likely to fly

Here is how ESA's future rocket, the Ariane 6, will appear in its four-booster configuration, delivering heavy satellites into Earth orbit.

cosmonauts on missions beyond Earth orbit, including to set up a base on the Moon by 2030. The central rocket, with four boosters attached, will lift communications satellites and other heavy payloads before then.

Europe has agreed to build its own powerful Ariane 6 rocket, building on the success of the reliable Ariane 5. This powerful new three-stage rocket will fly with either four boosters or two boosters, with the aim of keeping ESA competitive in the satellite-launching business, and is scheduled to make its first flight from Kourou, French Guiana, in 2020. The two-booster version will be the A62 and the four-booster the A64.

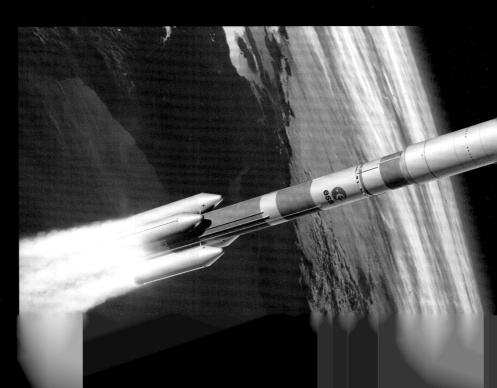

COMMERCIAL SPACEFLIGHT

Though NASA still leads the way in US space exploration, the market has been opened up to commercial companies, which are developing their own rockets and spacecraft. The space agency has so far allowed two of them, SpaceX and Orbital ATK, to fly cargo missions to the International Space Station. Several have been successfully flown by SpaceX's Dragon cargo craft and Orbital's equivalent, Cygnus.

SpaceX is also developing a crewed version of their capsule, called Dragon V2, that is due to carry astronauts to the ISS in 2017, ending the Americans' reliance on Russian Soyuz craft to reach the orbiting outpost. A Dragon was sent to the ISS in April 2016 with a test model of an inflatable module called Beam, built by Bigelow Aerospace, another commercial company that aims in time to have its own space station constructed from its expandable modules.

SpaceX launch their missions to the ISS on a Falcon 9, one of their own family of rockets. It is a simple, two-stage launcher with the first stage powered by nine engines to increase reliability. The company has begun tests to make their rockets reusable by bringing the first stage back to a vertical soft landing after launch.

A bigger rocket, the Falcon Heavy, is based on the Falcon 9 but has two extra nine-engined boosters attached, allowing it to fly the heaviest of payloads into space. SpaceX founder Elon Musk says his ultimate aim is to fly missions to Mars.

▶ Boeing's design for a crew transporter is the CST-100 Starliner, a capsule that will ferry astronauts to orbit in comfort.

▼ Resembling a smaller Space Shuttle, the SNC's Dream Chaser is shown after returning from orbit in this artist's impression.

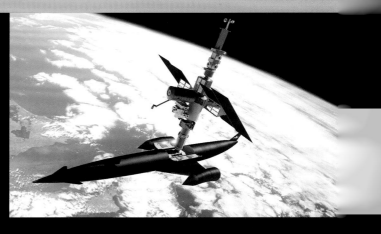

▲ A revolutionary spacecraft design is the UK's Skylon, which will be able to fly directly from runway into orbit, imagined here on a mission to a future space station.

Orbital ATK's Cygnus spacecraft is launched by a medium-sized rocket called Antares that the company built itself, either from Cape Canaveral, Florida, or NASA's Wallops Flight Facility in Virginia. It is another two-stage rocket, but a third stage is offered to reach higher orbits.

After three successful flights to the ISS in 2013 and 2014, the Antares suffered a catastrophic explosion seconds after launch of its fourth cargo mission on 28 October 2014. Two Cygnus supply missions subsequently flew on replacement Atlas V rockets. Flights of the Antares were due to resume in July 2016.

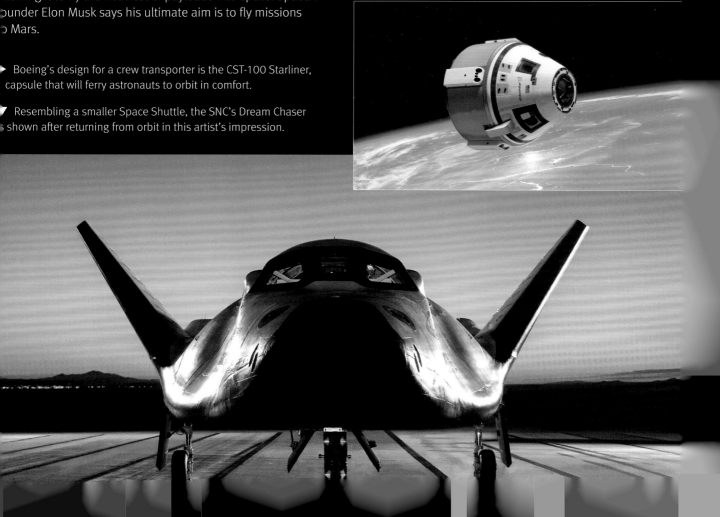

▲ Artwork showing SpaceX's Dragon V2 which will allow the United States to fly astronauts to the International Space Station once again after relying on Russian rockets for several years.

A separate crewed spacecraft, the CST-100 Starliner, is being developed by Boeing as an automated 'space taxi' that could transport astronauts on the short flights necessary to reach the ISS. Its first crewed flight is expected before 2020. The spacecraft is designed to be adaptable for a number of different launchers.

Another spacecraft being developed to fly into low-Earth orbit is the Dream Chaser, which resembles the Space Shuttle, and will be launched with its wings folded atop a rocket before returning to land on a runway. The Sierra Nevada Corporation, based in Colorado, which built the vehicle, has been awarded a contract by NASA to deliver and bring back cargo from the ISS on a number of unmanned flights, but the Dream Chaser is designed to be able to carry astronaut crews too.

▶ Orbital ATK's Antares rocket suffers a catastrophic explosion moments after launch in October 2014, showing that spaceflight is never routine.

FROM RUNWAY INTO SPACE

A different approach is being adopted by UK company Reaction Engines, which has designed an 82 m long spacecraft called Skylon that would take off from a runway to fly payloads directly into orbit. This is impossible using regular technology due to the amount of fuel that would need to be carried. But the company has designed a revolutionary hybrid Synergetic Air-Breathing Rocket Engine (SABRE), which mixes hydrogen with air drawn in from the atmosphere before switching to rocket mode. The technology, which will allow a swift turnaround between flights to space stations, has attracted interest from the US military as well as ESA.

SPACE TOURISM

PACK YOUR BAGS – YOU'RE GOING INTO SPACE!

Until now anyone wanting to fly in space has had to train to be an astronaut with an established agency. A small handful of tycoons have paid a lot of money to spend a few days on the International Space Station, but for the vast majority of people, going into orbit remains out of reach.

That is set to change. Some commercial companies are developing spacecraft that will allow ordinary folk to take a trip to the final frontier. To begin with, they will still need to have deep pockets, but the cost of a spaceflight should come down considerably in the future. The G-forces on the body of a suborbital flight are bearable by a moderately fit person, and the prospect is that space travel will indeed become available to the masses.

SPACE ADVENTURES

The world's first space tourists flew established routes into space because their 'holiday company', Space Adventures, booked seats on Russian Soyuz spacecraft that were carrying crews to the ISS. With tickets costing around $20 million a time, this has been an adventure that only the richest could afford. So far, seven fare-paying passengers have spent several days each on the International Space Station, following months of training with other cosmonauts at Star City, Moscow. The Virginia-based company is promising to offer flights around the Moon in the future, aboard an adaptation of the Soyuz spacecraft, and again in collaboration with Russia's Roscosmos space agency.

HOTELS IN ORBIT

Though the first tourist ships will make brief flights into space, orbital voyages are sure to follow. Bigelow Aerospace, headed by hotels tycoon Robert Bigelow, is planning to provide accommodation for them in the shape of space stations formed of inflatable sections. The concept was successfully demonstrated using a test module attached to the International Space Station in May 2016.

The first orbiting hotels could be made up of inflatable modules, such as in this concept artwork by Bigelow Aerospace.

VIRGIN GALACTIC

Hundreds of thrillseekers have already put deposits down with Virgin Galactic, which will begin suborbital flights from a custom-built spaceport at Mojave, New Mexico. Many have been waiting several years, but Virgin have always said they will not fly until ready as safety is paramount.

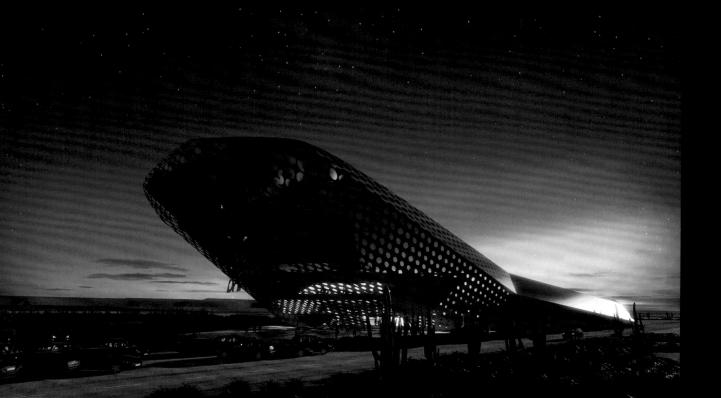

▲ The two-seater Lynx spaceplane which, because of its small size, will be able to fly to take off from a runway to make a suborbital flight before returning to land.

For $250,000, they are promised an adventure where their spaceship, carrying six passengers and a crew of two pilots, will be hoisted aloft by a twin-hulled carrier aircraft, called WhiteKnightTwo, to an altitude of 15,000 m. At that point it will be released and a single rocket engine will propel them above the Kármán line, 100 km high, where space officially begins. For a few minutes they can experience weightlessness and enjoy views of the curvature of the Earth before the spacecraft glides back to land on the spaceport's runway.

A tragic accident in October 2014 saw the break-up of Virgin's spacecraft, SpaceShipTwo, killing one of its two test pilots and severely injuring the other. The crash set the project back further, and by mid-2016 there was still no firm schedule for commercial operations.

BLUE ORIGIN

Blue Origin, headed by Amazon founder Jeff Bezos, is developing a two-part spacecraft called New Shepard. This 18 m high, single-stage rocket will lift off with a crewed capsule on top, carrying six passengers above the Kármán line to spend several minutes in weightlessness and enjoying the views. Meanwhile, the rocket will return to land itself vertically, ready for reuse, while the capsule parachutes to the ground. Like SpaceX with its Falcon 9, Blue Origin has successfully demonstrated the ability to land its rocket after a mission.

◀ An impression of the imposing spaceport that will welcome trippers taking suborbital flights from Curaçao in the Caribbean.

▶ Blue Origin's New Shepard's single stage returns to land itself after a test launch. A passenger capsule will parachute back to Earth.

LYNX SPACEPLANE

A more intimate experience is being offered by XCOR Aerospace, using the Lynx spaceplane that is being developed for flights out of spaceports including Mojave, California, and the Caribbean island of Curaçao. The vehicle can carry two people – a pilot and one passenger – on a half-hour suborbital flight, costing around $100,000, that gives five to six minutes in space. The spacecraft is designed to take off and land on a runway and will also be able to carry satellites into space.

COLONIES ON OTHER WORLDS

RETURN TO THE MOON

The success of the Apollo missions could have been the start of regular human visits to the Moon. Instead, there was an attitude that the challenge to get there had been met and there was no need to continue.

More than 40 years later, there is at last talk about returning to the Moon, with American, Russian and Chinese space scientists suggesting plans to go there. The focus in future will not be just to land for a short stay, but instead to set up lunar bases and possibly a lasting presence on the Moon similar to the continual stay by astronauts on the International Space Station.

Since Apollo, it has been discovered that the Moon has vast stores of water ice in its polar craters. Future missions to the Moon are likely to attempt to draw upon these supplies. They could provide not only water for astronauts to drink, wash and use for other purposes, but as a resource from which hydrogen fuel and oxygen can be extracted.

NASA and ESA have forward planning departments looking for new ideas to support space exploration. Those for a lunar base include delivering skeleton structures for habitats that robotic 3D printers can then cover with layers of lunar soil, or regolith, to make them stronger and provide insulation. Nuclear generators and solar panels could supply power. In the long term, commercial mining of the water ice would allow gas stations to be set up in orbit to fuel spacecraft.

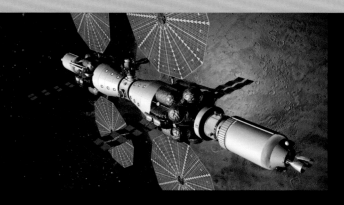

▲ A concept for an orbiting outpost at Mars to prepare the way for human landings on the planet, designed by aerospace company Lockheed Martin.

TO MARS AND BEYOND

The Moon may be a convenient stepping stone to more remote parts of the Solar System, but there are many who are impatient to get to Mars. Many people have volunteered for one-way missions to the Red Planet, though the nine-month voyage there will be challenging and plans remain on the drawing board. Others are experimenting by simulating long-term stays in deserts on Earth.

By the 2030s, the first steps towards a Mars colony may have been taken, either by sending astronauts in NASA's Orion spacecraft, or as part of SpaceX's aim to fly there. What is certain is that to set up a human presence on Mars, a considerable infrastructure will need to be prepared in advance – it is not

▼ ESA's concept for a Moonbase uses 3D-printing robots to cover a dome with lunar soil to protect it from extreme temperatures and meteoroids.

▲◄ A NASA artist imagines how humans might explore Mars (above), and a concept for a Martian greenhouse (left), where they could grow vegetables for fresh food.

MINING ASTEROIDS

NASA has declared its aim to send astronauts to a passing asteroid, to learn more about them and how to deflect any that threaten to collide with Earth in the future. There is also huge commercial interest because of the value of an asteroid's precious resources. Experts estimate that even a small asteroid could contain trillions of dollars worth of material, from water for rocket fuel to rare metals such as platinum that are essential for mobile phones, medical devices and electronic gadgets, and neodymium needed for the powerful magnets in electric cars. Excavating them from asteroids will be more environmentally friendly than mining sensitive regions of the Earth.

right on our doorstep like the Moon. Robotic machines will be sent by earlier spacecraft to deliver habitation modules, fuel generators and mechanisms to collect water from the Martian subsurface and extract oxygen from the atmosphere. Meanwhile, early astronauts may stay in orbiting space stations.

Other experiments are being carried out to see what sort of vegetables might grow best in the Martian soil, so that the colonists can be self-sufficient rather like the hero of the hit novel and movie *The Martian*.

Reaching Mars is not enough for some far-thinking individuals. A group called Objective Europa has been set up, based in Denmark, to harness the expertise of a wide range of people via the internet to plan for a crewed mission to Jupiter's ice moon Europa, which has been found to have a vast underground ocean. Organizers say such a trip is achievable using current technology and could include a submarine that would melt its way through the surface ice to explore the subterranean sea!

▶ Mining company Deep Space Industries imagines how a futuristic spacecraft might harvest the valuable minerals that are found inside an asteroid.

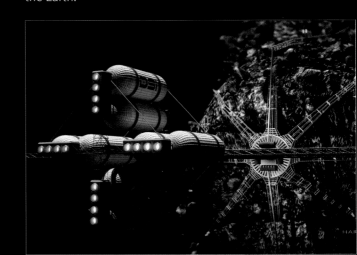

CAN SCIENCE FICTION BECOME SCIENCE FACT?

Space travel is still at an early stage and does not resemble the fantastic interstellar missions we see on TV and in the movies. Shows such as *Star Wars*, *Star Trek* and *Doctor Who* inspire legions of fans, but are we ever likely to see missions to explore nearby stars, let alone the rest of our own and other galaxies? So far, two spacecraft, the Voyager 1 and 2 probes, are at the boundary of the Solar System, nearly 40 years after they launched. But it would take them 80,000 years to reach even the nearest star, Proxima Centauri, 4.25 light years away, or more than 40 trillion km distant.

New technologies will need to be developed if humans are to make such fantastic voyages. And scientists and engineers have already been busy planning such missions – in fact, the first serious studies began half a century ago!

The United States' defence department's research wing, DARPA, is backing an initiative called 100 Year Starship that brings serious brains together, from philosophers to engineers, to design a mission to fly humans to another star system.

The study resembles Project Daedalus, conceived by experts within the British Interplanetary Society in the 1970s, to build a huge rocket in orbit. Powered by nuclear fusion,

◄ Two Icarus starships, Starfinder and Pathfinder, pause at Jupiter to fuel up on hydrogen from its atmosphere before beginning their interstellar voyages in this artist's concept.

it would then fly to a nearby star, accelerating to a sixth the speed of light. Even so, the journey would still take 50 years.

The project was superseded by Icarus Interstellar, which seeks to make such a mission happen by the end of the century. The first such flight may well be purely robotic, controlled by its onboard computers, and sending home data about the star system it explores – data that will take more than four years to reach us at the speed of light.

A crewed mission would be something far beyond even sending humans to Mars, which is right on our doorstep by comparison. Unless 'warp-drive' style technology could be developed for real to send spacecraft at speeds approaching that of light, the voyage to a rocky planet within a distance of 50 light years is certain to be a multi-generation one, with astronauts being born, starting families, and dying on the way there. The alternative of putting a crew into hibernation, as seen in sci-fi movies, is still just a concept.

Such incredible voyages may sound like fantasy today. But not so long ago, so did any idea of travelling in space!

SHIELDS UP!

Another idea from *Star Trek* is the shield that deflects attack by enemy spaceships. Though not on such a scale, researchers are developing a real shield for spacecraft that would protect crews from potentially deadly levels of natural radiation on long voyages. A UK-led team has found a way to create a barrier by mimicking the natural force field that protects the Earth from space weather – its magnetosphere. A small prototype has been seen to work in a laboratory, and studies are now continuing into how to scale the technology up to protect astronauts on interplanetary missions.

▲ Though still in the realms of science fiction, some believe spacecraft of the future will use wormholes to take shortcuts across space and time.

TRAVEL IN TIME AND SPACE

Another way that some argue is theoretically possible to cross the Universe is via wormholes – shortcuts between different points in space-time, the continuum that brings space and time together in physics. Such links were suggested by the brilliant physicist Albert Einstein and are said to be supported by his general theory of relativity. But no one knows whether they exist or what they would really be like, and many doubt that a real life TARDIS would be able to survive the extreme forces within one.

▼▶ Travel to the stars may still be a dream, but could travel posters like these from NASA tempt voyagers to make the trip one day?

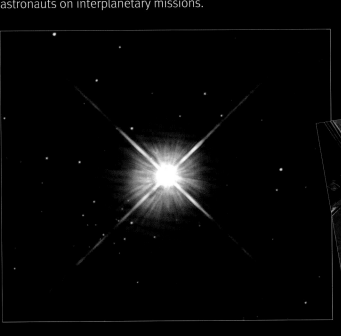

▲ Proxima Centauri, the closest star after the Sun, shines brightly in an image taken with the Hubble Space Telescope. Will humans be heading there this century?

INDEX

ACKNOWLEDGEMENTS

I offer grateful thanks to Megan Argo, Emily Baldwin, Amanda Doyle, Lucie Green, Lucy Rogers and Robin Scagell for their helpful advice and expertise during the writing of this book, to my mother Beryl Sutherland for proof-reading and offering valuable suggestions, to Chris Bell for her beautiful designs, and to Caroline Rayner of Philip's for making the project such a pleasure to work on.

1 NASA/JPL-Caltech/SETI Institute; 3 NASA; 5tl NASA, ESA, Hubble Heritage, S. Van Dyk (JPL/IPAC), R. Chandar (U. Toledo), D. De Martin and R. Gendler; 5tr NASA/JPL/USGS; 5b NASA; 6tl B.A.E. Inc./Alamy Stock Photo; 6tr NASA; 6b NASA, ESA, and M. Kornmesser; 7l Ed Rhodes/Alamy Stock Photo; 7r NASA/JPL-Caltech; 8 NASA/JPL-Caltech/UCLA; 10–11 NASA, ESA, Hubble Heritage, S. Van Dyk (JPL/IPAC), R. Chandar (U. Toledo), D. De Martin and R. Gendler; 12, 13t NASA/WMAP Science Team; 13b NASA, ESA, H. Teplitz and M. Rafelski (IPAC/Caltech), A. Koekemoer (STScI), R. Windhorst (Arizona State University), and Z. Levay (STScI); 14 NASA/JPL-Caltech; 15t NASA/ESA and The Hubble Heritage Team (STScI/AURA); 15b NASA, ESA, Z. Levay and R. van der Marel (STScI), T. Hallas, and A. Mellinger; 16 NASA/JPL; 17t NASA, ESA, the Hubble Heritage Team (STScI/AURA), A. Nota (ESA/STScI), and the Westerlund 2 Science Team; 17b NASA, ESA, and A. Schaller (for STScI); 18–19, 20 NASA; 21t Detlev Van Ravensway/Science Photo Library; 21c Félix Nadar (Public Domain); 21b SPUTNIK/Science Photo Library; 22 Alexander Gordeyev/Dreamstime.com; 23tl ESA – Pierre Carril; 23tr Philip's; 23b, 24t NASA; 24b NASA/JPL-Caltech; 25t NASA/Johns Hopkins University Applied Physics Laboratory/Southwest Research Institute (JHUAPL/SwRI); 25b ESA/NASA; 26 Copyright 2006 EUMETSAT; 27t ESA/D. Ducros; 27b NASA/JPL-Caltech; 28t ESA, image by AOES Medialab; 28b NASA/JPL-Caltech; 29t NASA Johnson Space Center; 29c NASA/JPL-Caltech/MSSS; 29b NASA/WMAP Science Team; 30 RGB Ventures/SuperStock/Alamy Stock Photo; 31t NASA; 31c NASA/JPL-Caltech; 31b NASA; 32t, 32b SPUTNIK/Alamy Stock Photo; 33t Everett Collection Historical/Alamy Stock Photo; 33bl Granger, NYC/Alamy Stock Photo; 33br NASA; 34–35 NASA/JPL/USGS; 36, 37t, 37c, 37b, 38t, 38b, 39t, 39b, 40t, 40b, 41t, 41c, 41b, 42t, 42c, 42b, 43l, 43r NASA; 44t Russell L. Schweickart; 44b NASA/David Scott; 45t, 45b NASA; 46 NASA/Neil A. Armstrong; 47t Nerthuz/Dreamstime.com; 47bl, 47br NASA; 48t, 48b NASA/Neil A. Armstrong; 49tl, 49tr NASA; 49c Alexkalina/Dreamstime.com; 49b NASA; 50t NASA/Charles Conrad Jr; 50b NASA/Richard Gordon; 51t NASA/Charles Conrad Jr; 51bl NASA/Apollo 12 crew; 51br, 52t NASA; 52c NASA/Apollo 13 crew; 52b, 53t, 53c, 53b, 54t, 54b NASA; 55t Edwin Verin/Dreamstime.com; 55b NASA/David R. Scott; 56t NASA; 56b NASA/John W. Young; 57t NASA/Harrison H. Schmitt; 57c, 57b NASA/Eugene Cernan; 58 NASA/Neil A. Armstrong; 59t NASA; 59b Alexander Perepelitsyn/Alamy Stock Photo; 60t Everett Collection Historical/Alamy Stock Photo; 60b, 61t, 61b NASA; 62–63 B.A.E. Inc./Alamy Stock Photo; 64 SOHO (ESA and NASA); 65t NASA/SDO/HMI; 65b WaterFrame/Alamy Stock Photo; 66t NASA/SDO/Wiessinger; 66b NASA; 67t SOHO (ESA and NASA); 67c Alexkalina/Dreamstime.com; 67bl SOHO (ESA and NASA); 67br NASA/TRACE; 68t ESA and NASA/SOHO; 68b ESA/NASA; 69t NASA; 69bl SOHO (ESA and NASA); 69br SST/Institute for Solar Physics/Göran Scharmer and Kai Langhans; 70 NASA/JPL/USGS; 71t NASA's Goddard Space Flight Center/SDO/Genna Duberstein; 71b, 72t NASA/Johns Hopkins University Applied Physics Laboratory/Carnegie Institution of Washington; 72b MESSENGER, NASA, JHU APL, CIW; 73t ESA – image by C. Carreau; 73cl Alexkalina/Dreamstime.com; 73cr, 73b NASA/Johns Hopkins University Applied Physics Laboratory/Carnegie Institution of Washington; 74 NASA/JPL; 75t NASA; 75b ESA – AOES Medialab; 76t ESA; 76bl ESA/NASA/JPL; 76br NASA/JPL; 77t EVE/T.Balint; 77c Alexkalina/Dreamstime.com; 77b NASA/JPL-Caltech; 78 NASA/GSFC/Arizona State University; 79t NASA/NOAA; 79b NASA/JPL-Caltech; 80, 81tl NASA/GSFC/Arizona State University; 81tr Alexkalina/Dreamstime.com; 81c NASA/JPL-Caltech/MIT/GSFC; 81b NASA/JPL; 82t Chinese National Space Administration, Xinhuanet; 82b ISRO/NASA/JPL-Caltech/Brown University/USGS; 82t epa european pressphoto agency b.v./Alamy Stock Photo; 83b Xinhua/Alamy Stock Photo; 84 NASA/JPL-Caltech; 85t NASA, ESA, the Hubble Heritage Team (STScI/AURA), J. Bell (ASU), and M. Wolff (Space Science Institute); 85b NASA/JPL/University of Arizona; 86t ESA/DLR/FU Berlin; 86c Alexkalina/Dreamstime.com; 86–87b NASA; 87t, 87cr NASA/JPL-Caltech/University of Arizona; 87cl NSSDC/NASA; 88t NASA/GSFC; 88b ESA/DLR/FU Berlin (G. Neukum); 89l NASA's Goddard Space Flight Center; 89r ESA/DLR/FU Berlin; 90t NASA/JPL-Caltech/Cornell/USGS; 90b NASA/JPL-Caltech/MSSS; 91t ESA – AOES Medialab; 91b NASA/JPL-Caltech/MSSS; 92t Processed image by Justin Cowart, The Planetary Society, Original data by NASA/ JPL/California Institute of Technology; 92b NASA/JPL-Caltech; 93t NASA/JPL-Caltech/UCLA/MPS/DLR/IDA; 93b NASA/JPL-Caltech/UCLA/MPS/DLR/IDA/PSI/LPI; 94t NASA/JPL-Caltech/UCLA/MPS/DLR/IDA/PSI; 94b NASA/Goddard/University of Arizona; 95t JAXA/Akihiro Ikeshita; 95c NASA/JPL/USGS; 95b NASA/JPL-Caltech/JAXA/ESA/Galaxy; 96–97 NASA; 98 ITAR-TASS Photo Agency/Alamy Stock Photo; 99t, 99c NASA; 99bl SPUTNIK/Alamy Stock Photo; 99br, 100, 101t NASA; 101c Alexkalina/Dreamstime.com; 101b, 102 NASA; 103t Nerthuz/Dreamstime.com; 103c NASA; 103bl SPUTNIK/Alamy Stock Photo; 103br Alexkalina/Dreamstime.com; 104t, 104b, 105cl, 105cr, 105b, 106t, 106b, 107l, 107r, 108t NASA; 108b ESA/NASA; 109t NASA; 109c Alexkalina/Dreamstime.com; 109b, 110t, 110b, 111t, 111c, 111b NASA; 112 NASA/Tracy Caldwell Dyson; 113t ESA/NASA; 113c, 113b, 114t, 114b NASA; 115t NASA/Bill Ingalls; 115b NASA; 116, 117t, 117bl Corbis; 117br NASA; 118–119 NASA, ESA, and M. Kornmesser; 120 NASA Ames; 121tl NASA/JPL/USGS; 121tr NASA's Goddard Space Flight Center; 121b, 122t, 122b NASA/JPL; 123t Alexkalina/Dreamstime.com; 123c NASA/JPL; 123b NASA; 124 NASA/JPL/Space Science Institute; 125t NASA; 125c Alexkalina/Dreamstime.com; 125b Paul Sutherland; 126t, 126b NASA; 127t Hubble Space Telescope Comet Team and NASA/ESA; 127b NASA/ESA/K. Retherford/SWRI; 128t Galileo Project, SSI, University of Arizona, JPL, NASA; 128b NASA/JPL-Caltech/SETI Institute; 129t NASA/JPL/DLR; 129b NASA/JPL-Caltech; 130t NASA/JPL/DLR; 130b NASA/JPL/Ted Stryk; 131t NASA; 131c NASA, ESA and E. Karkoschka (University of Arizona); 131b NASA/JPL/Cornell University; 132t NASA/JPL/University of Arizona; 132b NASA/Johns Hopkins University Applied Physics Laboratory/Southwest Research Institute/Goddard Space Flight Center; 133t ESA/AOES; 133b NASA/JPL-Caltech; 134 NASA/JPL/Space Science Institute; 135t Cassini Imaging Team, SSI, JPL, ESA, NASA; 135b Alexkalina/Dreamstime.com; 136 NASA/JPL; 137t NASA/JPL-Caltech/Space Science Institute/G. Ugarkovic; 137c, 137b NASA/JPL/University of Colorado; 138t NASA/JPL/University of Arizona; 138b NASA/JPL-Caltech; 139t NASA/JPL; 139c NASA/JPL-Caltech/SSI/Hampton University; 139bl NASA; 139br Alexkalina/Dreamstime.com; 140t NASA/JPL-Caltech/Space Science Institute; 140b NASA/JPL-Caltech/University of Arizona/University of Idaho; 141tl NASA/JPL-Caltech; 141tr ESA/NASA/University of Arizona; 141b NASA/JPL/ESA; 142t NASA/JPL-Caltech/Space Science Institute; 142b NASA/JPL-Caltech; 143tl NASA/JPL; 143tc NASA/JPL-Caltech/Space Science Institute; 143tr NASA/JPL/Space Science Institute; 143bl NASA/JPL-Caltech/Space Science Institute; 143br NASA/JPL/Space Science Institute; 144 NASA/JPL-Caltech; 145tl, 145tr, 145tcl, 145tcr NASA/JPL; 145bcl NASA/JPL-Caltech; 145bcr NASA/JPL; 145bl NASA/E. Karkoschka/University of Arizona; 145br Alexkalina/Dreamstime.com; 146t, 146b NASA; 147t NASA/JPL/USGS; 147c Alexkalina/Dreamstime.com; 147b NASA/JPL; 148, 149t, 149b, 150t NASA/Johns Hopkins University Applied Physics Laboratory/Southwest Research Institute; 150c Alexkalina/Dreamstime.com; 150b NASA/Bill Ingalls; 151t, 151c, 151b NASA/Johns Hopkins University Applied Physics Laboratory/Southwest Research Institute; 152 NASA/JPL-Caltech; 153tl NASA and G. Bacon (STScI); 153tr NASA, ESA and Adolf Schaller; 153b NASA, ESA, and G. Bacon (STScI); 154 NASA/JPL-Caltech/UMD; 155t Paul Sutherland; 155b ESO/S. Deiries; 156t ESA/Rosetta/MPS for OSIRIS Team MPS/UPD/LAM/IAA/SSO/INTA/UPM/DASP/IDA; 156b ESA/ATG medialab, Comet image: ESA/Rosetta/NavCam; 157t ESA – Jürgen Mai; 157bl ESA/Rosetta/MPS for OSIRIS Team MPS/UPD/LAM/IAA/SSO/INTA/UPM/DASP/IDA; 157br Alexkalina/Dreamstime.com; 158–159 Ed Rhodes/Alamy Stock Photo; 160t Reinhold Wittich/Dreamstime.com; 160b, 161tl Public Domain; 161tr Michael Jenner/Alamy Stock Photo; 161b Paul Sutherland; 162t Alex Gorodnitchev/Dreamstime.com; 162b Inge Hogenbijl/ Dreamstime.com; 163t ESO/C. Malin; 163b Dan Talson/ Dreamstime.com; 164t SKA Organization; 164b ESO/L. Calçada; 165t Giant Magellan Telescope – GMTO Corporation; 165c NASA; 165b Todd Mason, Mason Productions Inc./LSST Corporation; 166t NASA, ESA, and The Hubble Heritage Team (STScI/AURA); 166b ESA and the Planck Collaboration; 167t NASA/CXC/MIT/UMass Amherst/M.D.Stage et al.; 167c Designua/Dreamstime.com; 167bl NASA/Swift/Stefan Immler (GSFC) and Erin Grand (UMCP); 167br ESA/C. Carreau; 168t, 168b NASA/ESA; 169t The Hubble Heritage Team, (STScI/AURA), ESA, NASA; 169bl NASA, ESA, and The Hubble Heritage Team (AURA/STScI); 169bcl NASA, ESA and the Hubble SM4 ERO Team; 169bcr NASA, ESA, Hubble Heritage Team; 169br NASA, Holland Ford (JHU), the ACS Science Team and ESA; 170t, 170c NASA Ames/JPL-Caltech/T. Pyle; 170b NASA Ames/JPL-Caltech/T. Pyle; 171t NASA Ames/JPL-Caltech/T. Pyle; 171b NASA/Goddard Space Flight Center; 172t ESA (Image by AOES Medialab) – background: Hubble Space Telescope, NASA/ESA/STScI; 172c NASA/CXC/NGST; 172b X-ray: NASA/CXC/U.Texas/S.Post et al, Infrared: 2MASS/UMass/IPAC-Caltech/NASA/NSF; 173t ESA/ATG medialab – background: ESO/S. Brunier; 173b NASA/JPL-Caltech; 174t, 174c NASA; 174b NASA/Northrop Grumman; 175t, 175b NASA; 176–177 NASA/JPL-Caltech; 178, 179tl NASA; 179tr NASA/MSFC; 179b ESA/D. Ducros; 180t Courtesy Reaction Engines Ltd; 180c Copyright © 2014 The Boeing Company. All Rights Reserved.; 180b NASA; 181t SpaceX; 181b NASA/Joel Kowsky; 182t Courtesy Bigelow Aerospace; 182b, 183t Courtesy XCOR Space Expeditions; 183b Blue Origin/Alamy Stock Photo; 184t Courtesy Lockheed Martin; 184b ESA/Foster + Partners; 185t NASA/Pat Rawlings, SAIC; 185c NASA; 185b Courtesy Deep Space Industries/Bryan Versteeg; 186 Adrian Mann; 187t NASA/Glenn Research Center (digital art by Les Bossinas (Cortez III Service Corp.), 1998); 187bl ESA/Hubble and NASA; 187bc, 187br NASA/JPL-Caltech.

Front endpaper (Dust pillars in the Carina Nebula): NASA, ESA, and the Hubble Heritage Project (STScI/AURA).

Rear endpaper ('Mystic Mountain', Carina Nebula): NASA, ESA and M. Livio and the Hubble 20th Anniversary Team (STScI).

Front cover: tl NASA/ESA/K. Retherford/SWRI; tc NASA and The Hubble Heritage Team (STScI/AURA); tr NASA/Johns Hopkins University Applied Physics Laboratory/Southwest Research Institute/Steve Gribben; c NASA/TRACE; bl NASA; bc NASA/JPL-Caltech/SETI Institute; br NASA.

Back cover: cl NASA/JPL-Caltech; c NASA/JPL-Caltech; cr NASA, ESA, the Hubble Heritage Team (STScI/AURA), A. Nota (ESA/STScI), and the Westerlund 2 Science Team; b NASA/JPL-Caltech.